HANDCRAFTED
DOORS & WINDOWS

HANDCRAFTED
DOORS & WINDOWS

Written by Amy Zaffarano Rowland
Designed by Barbara Field

Rodale Press, Emmaus, Pennsylvania

Printed in the United States of America.

Book Layout by John Pepper
Creative Development of Artwork by Keith and Linda Heberling
Camera-Ready Art by Frank Rohrbach

The doors in the front cover photo are by Didrik Pederson; those shown on the back cover are by Victor Hiles (top photo) and Louis Villarreale (bottom photo). The windows on the front cover are by Carla Wingate; those on the back cover are by Adele Hiles (top) and John Forbes (bottom).

Library of Congress Cataloging in Publication Data

Rowland, Amy Zaffarano.
    Handcrafted doors and windows.

    Bibliography: p.
    Includes index.
    1. Doors. 2. Windows. 3. Woodwork. I. Title.
TH2278.R65   1982   694'.6   82-11252
ISBN 0-87857-423-9 hardcover
ISBN 0-87857-424-7 paperback

  4 6 8 10 9 7 5 3   hardcover

  4 6 8 10 9 7 5   paperback

CONTENTS

PREFACE

It is easy to appreciate a work of craftsmanship from an earlier time. Whatever other merit the work has, it has endured—and that is a measure of the care which was taken in its construction. The 200-year-old post-and-beam barn that still stands, the turn-of-the-century wooden cradle discovered in the attic, the heirloom quilt passed from mother to daughter for generations—these artifacts of other, earlier lives become part of our own because they were made by people who intended them to *last*. So they used the best materials available to them at the time, worked with patient skill and high standards, and strove for beautiful design. Although these people thought of themselves as quite ordinary, with the perspective of hindsight, we call them craftsmen.

They left us a legacy of objects necessary for daily life—cast iron pots, handwoven table covers, ladderback chairs, wheel-

thrown crockery bowls, and the like. And in doing so, they defined in another way the traditional concern of the craftsman to create objects that serve a practical purpose—objects that we use to gather and serve our food, to clothe our bodies, and to build and furnish our houses.

Of course, the way in which we feed, clothe and shelter ourselves is just as much a concern today as it was for our ancestors. Manufacturing companies mass-produce every item we can conceive of to meet our needs, usually striving not for quality or durability but for the most cost-effective way to meet demand. Because of this, many of us think of craftsmanship as a lost inheritance, a forgotten value we can be reminded of only by touring restored Colonial villages like Williamsburg, Virginia, or haunting country antique shops.

Yet there *are* contemporary craftsmen who work with just as much patience and care as the millwrights and joiners of 200 years ago and the glass artisans at the turn of the century. Their work is on display in the hundreds of craft fairs that now take place in the United States each year, in galleries in every major city, and in exclusive museums. It is photographed for the pages of glossy magazines like *American Craft, Glass Studio,* and *Fine Woodworking.* And it is taught at crafts schools around the country.

Although some contemporary craftsmen concern themselves more with form than with function, the majority still create objects that are intended for use in daily life—music stands, bedsteads, desks, weather vanes, serving bowls, utensils, vests, shawls, jewelry. And they make these practical objects with fine materials and by hand, in the tradition of our ancestors. For the most part, their designs are not historical reproductions but originals, and take advantage of the precisely machined tools and equipment and carefully formulated glues used by modern manufacturers.

In turn, manufacturers are influenced by craftsmen's innovative designs. But manufacturers are not the only ones! Ordinary people who have learned to appreciate the value of craftsmanship try, in various ways, to incorporate it and express it in their own lives.

It is perhaps no coincidence that the owner/builder movement has grown strong during the same period of time that has seen a resurgence of interest in crafts. The two are linked: Both the owner/builder and the craftsman reclaim traditional values of quality in materials and workmanship. Both apply their skills to create something at once practical and pleasing. Is it any wonder then that craftsmen and owner/builders often work side by side, sharing their knowledge as well as their purpose?

Books like *Handmade Houses* and *Made with Oak* have documented the work of craftsmen/builders, but have let their work methods remain a mystery—with good reason. What makes a handcrafted house so special is not only the quality of materials used and the level

of workmanship attained, but also the unique personality of its builder, expressed in every detail of the design.

The photographs in *Handcrafted Doors and Windows* celebrate this originality. They say, in hundreds of ways, that making something beautiful is a joy, and acquiring the courage to be creative is exhilarating—an experience, in fact, you might like to try yourself.

If you value craftsmanship and want to express this value in your environment, you will find that making a door or window is a good way to begin. It demands thoughtful choices; it utilizes and improves practical skills; and it encourages creative design. *Handcrafted Doors and Windows* provides the information and inspiration necessary to guide you through this process, and make it a pleasure.

The book opens with an introductory chapter written by an award-winning architect and craftsman, Tom Bender. He discusses the meanings doors and windows have come to have in various cultures, and changes that have occurred as technology has made improved design the common property of aristocrats and the lower classes alike.

Part I, "The Making of Doors," covers the construction and design basics of door building. The first chapter, on structure, provides the woodworking terminology necessary for further discussion. The second chapter explains how to make your own door—batten, panel or flush—and then frame and hang it. The last chapter describes how to finish off both interior and exterior doors for beauty and durability, and how to choose hardware for security, fire safety and ease of access.

Part II, "A Gallery of Handcrafted Doors," takes you into the woodworking studios of ten doormakers to hear the stories of how they make doors and door hardware. You'll be surprised by how some of these men and women have come to be craftsmen, how little they often have in the way of tools and a workshop, and how differently they all work.

Part III, "The Making of Windows," discusses the essentials of sound window structure and the purpose served by various window forms—fixed, casement, double-hung, horizontal sliding, awning and others, and provides instruction on how to make and glaze your own windows, frame and install them.

"A Gallery of Handcrafted Windows," Part IV, profiles seven craftsmen and custom builders who have made unusual and striking windows or window accessories. Like the previous parts, it contains many photographs, and illustrations which clarify technical details.

Comprehensive appendices round out this format to provide resources for the homeowner who wants to try making his own door or window for the first time, or for the more experienced woodworker or glass artisan who needs hard-to-find materials and tools to carry out a special design. In addition, architects or homeowners who want to commission a door or window from a craftsman will find the directory of doormakers and windowmakers useful.

Handcrafted Doors and Windows is for those who appreciate craftsmanship and who aspire to it themselves.

ACKNOWLEDGMENTS

This book is the product of a tremendous amount of sharing on the part of many people. Carol Stoner, Editorial Director of Book Division, came up with the original idea for a book about handcrafted doors and windows. She and Bill Hylton, Managing Editor; Karen Schell, Art Director; and Tom Gettings, Photo Director, were strongly supportive throughout all phases of this project.

As designer of *Handcrafted Doors and Windows*, Barbara Field is largely responsible for making this book the beauty that it is. Her strong design sense eased the most monumental task of all—selecting photos for the book from the 2,600 submitted to us by free-lancers and the hundreds more taken by Rodale photographers.

Researching sites that Rodale might photograph and craftsmen and owner/builders whose work might be appropriate for this book put me in touch with many helpful people. I would particularly like to thank John Stolz, Jeff Boerner, Gino Russo, Jack Jarrell and the members of the Old Allentown Fairground Association for letting me tour their homes, and Lloyd Kahn of Shelter Publications, *New Shelter* Art Director John Johanek, David Haupert at *Better Homes and Gardens*, Anita Rosenfeld of the Bucks County Art Alliance, and Art Espenet of the Bolinas County Craftsman's Guild for sharing their contacts.

My thanks to all those who submitted photographs for consideration, but especially to Bill French, Robin Rothstein, Roy Mullin and Michael Kanouff, who among them found dozens of the sites photographed in this book.

All of the Rodale staff photographers—

Chris Barone, Carl Doney, Tom Gettings, John Hamel, Mitch Mandel, Sally Ann Shenk, Margaret Smyser, and Christie Tito—deserve many thanks, too.

Tom Bender, who wrote the introduction to the book, and James Warfield, Associate Professor at the School of Architecture of the University of Illinois, willingly shared their photo archives with us.

The Planning Commission of the City of Reading allowed us to reprint some of the photographs used in their beautiful "Doors of Reading" poster. The Borough Council of Jim Thorpe gave us permission to take photographs of doors and windows in the Asa Packer Mansion, and Gary Michael Jones, Curator/Historian of the Philadelphia Old Town Historical Society, invited our photographer and book designer into his home to take photographs of the restored nineteenth-century windows there.

Still others provided props for some of our photographs. Ted Connelly of the Dean Company, Randy Gelzer of Driwood Mouldings, Newton Millham, blacksmith, Whitman Ball of Ball and Ball Hardware, and Merrilyn Street of Renovator's Supply generously allowed us to borrow stock items. Minwax and Glidden Coatings and Resins, a division of SCM Corporation, supplied us with samples of their stains. Edward Lineberger, president of Allen Hardware, not only lent us many pieces of door hardware to photograph, but also kindly allowed me to talk with him for hours about door hardware functions, style and security. Tom Walz and Phil Gehret, woodworkers on Rodale's staff, made some of the other props we required.

Paul Moser of Moser Brothers, Inc., and Luther Walp of Drums Sash and Door were similarly generous with interview time and permission to take photographs on the premises. Jeffrey Walbert of Walbert Lumber Company; Joseph Takacs of Anchor Building Supply; Bill Fetterman of American Sash Company; Bill McCann, Arthur Paolini and Dave Rehrig of Robbins Door and Sash; and Tom Glose of Glose's Stained Glass all kindly granted me interviews.

My special thanks to all the craftsmen who submitted notes on their construction process and allowed me to interview them for gallery articles and boxes: Tom Anderson, Tom Barr, Tom Bender, Lesta Bertoia, Aj Darby, Joe Devlin, Rhonda Dixon, Bruce Fink, Al Garvey, Larry Golden, Adi Hienzsch, Bob Jepperson, Lynn Kraft, James Parker, Bob Richardson, Bruce Sherman, Charlie Southard, Megan Timothy, and Alexander Weygers.

As with any big project, there are many "assistants" who do much of the most demanding work. Special thanks to John Pepper, Shirley Smith, Rene Grimes, Janet Vera, Barbara Erich, Marge Wieder, Dorothy Smickley and especially Bobbie Hartranft.

Finally, those who reviewed part or all of the manuscript before it went into production—Tom Bender, Edward Lineberger, James Warfield, Jim Eldon, Marilyn Hodges, Bill Hylton, Jane Sherman, and Carol Stoner—have done an important service, correcting inaccuracies and making the book more readable by their comments.

Thanks to all these people and the many more who helped me. I hope you find the book well crafted.

INTRODUCTION

The unexpected guest knocking on our door can change our lives; the view out a window of a tree, strong and rooted in earth as it endures the changes of the seasons, can inspire us daily. Doors and windows frame many of our memories—not only the departures and home-comings, but also the "ordi-nary" pleasures of a house that make us think of it ever after-ward as a place where we were happy—the window alcove where we played guitar, the porch where we sat and chatted with neighbors.

It would seem that a piece of glass is all there is to a window, and that any old slab of wood would make an ade-quate door. There is more to a window than a view, however, and more to a door than privacy. Both should be designed with the needs of those who will use them in mind, and then crafted with materials that will last.

Often, we walk up to the

front door of our own house carrying boxes or babies or bags of groceries that we have to juggle while we grope for the key. How welcome a convenient shelf for such burdens would be! Sadly, few doormakers have given that much thought to their design. More recognize the importance of an ample porch roof or at least an awning over the entrance door. After all, almost all of us have stood in a downpour at one time or another trying to rouse someone inside the house.

We greet strangers, friends and relatives at the door, and there we say goodbye. We need room both inside and out to hug and cry and laugh and linger without becoming entangled in other events or lives, or falling off too narrow a doorstep. We need a place to sit down and put on boots, to take off coats, to put dripping umbrellas. A vestibule or entry hall allows us to come and go or meet unexpected visitors without impinging on other activities within the house. And in cold climates, an entrance design that uses the doors of the entry hall as an "air lock" keeps the rest of the house warm as we make our guests welcome or bid them a safe journey.

Part of the magic of doors and windows is that they are in-between places—where the inside and outside worlds come together, where private and public meet, where the known and the unknown walk hand in hand. We can sit on our doorstep or lean out our window and be a part of all that's going on in the street, yet have the option of ducking back into the house. Or we can sit quietly and watch without having to take part. As we go through a door, we shift our

Doors and windows connect us to the world of which we are part—neighbors, city streets, country scenes.

The torii *of Japan marks our passage into another part of the forest without detracting from its beauty.*

attention from what we're leaving behind to that which waits for us beyond. When we glance out a window, we get a change of scene and perspective.

The in-between realm of the front door has blossomed into very special forms — among them the stone doorsteps of East Coast row houses, the front porches of midwestern Victorian homes, and the verandas of Japanese houses. The entrance to one of the most famous Zen gardens in Japan is actually a beautifully landscaped walkway that winds along two sides of the garden before it brings you into the garden itself. The long, quiet walk gives you the opportunity to compose yourself so you can appreciate the beauty and serenity of the garden when you finally reach it.

The styles of doors evolved by a culture reflect the concerns of its people through time. The massive city gates of India, made of huge crossed wooden beams studded with giant iron spikes, were designed to withstand the charge of war elephants. The *torii* of Japan, two vertical posts with a beam across their tops, let us know we are leaving one space and entering another without interfering with our sense of unity with the world through which we walk. The curtain of wooden beads used as a passageway door in Mediterranean countries gives some visual privacy, but encourages and allows room-to-room conversations between relatives and friends. And Dutch doors, whose top half can be opened while the bottom half is closed, let in light and fresh air while keeping out the chickens wandering in the yard.

Doorcraft has reached some extraordinary

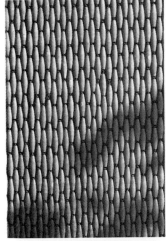

heights. France is home to many Gothic cathedrals, entered through doors that are masterpieces of wood and ornamental ironwork. Some of the fabled gold and silver doors of the ancient Persian mosques still exist, a reminder of the importance that people placed on their spiritual life. For many centuries, the tribes that cross the Hindustan Mountains have woven ornate tapestries that form the doors of their tents. Though less enduring, these have also come to be regarded as great works of art. Though we may never dream of equaling any of these masterpieces, they offer us as much inspiration as the wonderful carved redwood doors of California, or the simple batten doors of our Colonial period.

As important as the physical attributes of doors are their symbolic meanings and associations. In the Netherlands, almost every home has a freshly varnished wooden door and a doorstep scrubbed until it shines. Anything less is considered evidence of an inharmonious home. In Japan, most doors are delicate constructions of light wood and paper—frail barriers against

Doors serve many purposes.
Left, top to bottom: *A simple bead curtain filters the Mediterranean sunlight pouring into a villager's house. The handcarved door in Safed, Israel, reminds passers-by to honor all forms of life—the fruits of the earth and the creatures of the wild. The ornately decorated, heavy doors on the Cathedral of Notre Dame inspire awe of earthly wealth and power as well as glorifying God.*
Above: *Maria of San Ildefonso Pueblo, the most famous of the American Indian potters, bestows a rare favor on the photographer by emerging for a photograph.*

forced entry. However, the Japanese have imposed such a strong taboo against this act of violence that these slight doors are sufficient.

From time immemorial, the door has been the boundary of our protected and secure sources of food, shelter and warmth. Beyond lies the outside world, where all sorts of unknowns lurk. The door allows us to close out that world, and to relax the alertness that we need to deal with its challenges. When a door does not adequately separate public and private worlds, we become uncomfortable. Glass doors and picture windows beside doors can allow the approaching visitor to see into people's private lives to such an extent that he feels like an unwelcome intruder by the time he knocks on the door.

How a door is made conveys a lot about the hidden world beyond it. A grand entrance with a large panel door, elaborate polished brass hardware, and fanlight and pediment above it speaks of affluence, privilege and the cultivated taste of the occupants. A plain, sturdy door may open into a house whose tenants believe that simplicity of material things contributes to the richness of other aspects of their lives. How a door is made can show an appreciation of craftsmanship, a love of fine or rough-hewn materials, or a liking for what is salvaged or brand new.

While a door's primary task is to separate the house from the world outside, the main job of a window is to join together the inner and outer worlds. As well as providing light and ventilation, windows keep us in touch with the weather, the time of day, the changing seasons, the world of nature, and the lives of our neighbors. They draw our lives out of their narrow paths to touch the infinitely richer complexity and variety of life that goes on around us. In doing so, they help us keep a sense of the meaning of our lives in the world of which we are part.

The Japanese house, with its removable paper walls, appears at first inspection to merely have bigger windows or wider views. The practice of keeping the rooms open and unheated in the winter seems at best a hardship. But the very openness of the Japanese house allows more than a look into the garden. It makes the scent of spring, the silence of snow, the moon-

Above: *The bedroom window at the Asa Packer Mansion in Jim Thorpe, Pennsylvania, is just as it was when the steel magnate was alive, but the picture of life framed by the window changes.*
Opposite, top: *The tiny windows in the adobe buildings at Taos Pueblo shade those within from desert sun; the windows of a house in Finland are small enough to deflect the glare of sunlight on snow.*
Opposite, bottom: *In the more benevolent climate of the Swiss Alps, this medieval chalet opens itself up to sunshine and a spectacular view.*

rise, and the noonday quiet all ever-present parts of our consciousness.

A picture window, though it puts a barrier of glass between us and the outside world, also connects us to our environment. Through its wide glass, we can survey the city skyline or the trees on the mountain turning the color of fire with the spread of October's frosts. A window that frames a special view need not be large, however. A tiny window can focus our attention enough to see the wonder that exists in the opening of a single blossom, or the flashing rainbow of light transmitted by a single drop of dew on the point of a leaf.

Windows, far more than doors, have been modified to adapt to different climates. In overcast climates like that of England, big windows are common to keep rooms from being dark and depressing. There is rarely enough sun to overheat the rooms, and the heat lost to the outside is less significant than it would be in areas with severe winters. In Scandinavia, windows often shrink in size and huddle together on the sunny side of a house, both to conserve heat and because the dazzling light reflected off the snow-covered world can light a room easily through much smaller openings. And in the American Southwest, modern architects have taken a lesson from the pueblo builders, who designed their thick-walled adobe constructions with small windows to keep out the direct sun, making the rooms inside a cool and shady refuge from the desert heat.

In almost every climate, the sizing, placement and shading of windows have been instrumental to the success of passive solar buildings. Greenhouses, sunspace entrances, thermal chimneys, and simpler windows have been carefully placed to take advantage of the sun's light and

Above, top to bottom: *The Persians use delicate wooden grilles to diffuse the bright desert sunlight, and cupolas on the mosque dome roofs to help cool and light the interior.*

warmth. The further development and refinement of such windows is one of the challenges facing building designers today.

Diffusing sunlight rather than allowing it to penetrate directly into living spaces is a common way of coping with the strong sunlight near the equator. The Khmers of Cambodia use delicately carved wooden or stone spindles in window openings to scatter and soften the light, while giving privacy and security. The people of Venice, Italy, reduce the glare of bright sunlight reflecting off the canals by placing their windows at the corners of the room, so that the whole room is illuminated softly. A common way of contending with tropic heat and glare is to design rooms with windows on at least two sides, so that lighting is even and cross-ventilation is possible.

Adjustable louvered window shutters are used to provide shade and encourage natural ventilation in many hot climates. Below the Mason-Dixon line, many buildings are designed with high ceilings and tall, operable windows to allow the hot air to rise and escape from the rooms.

The seventeenth- and eighteenth-century palace buildings of India and Persia show how creatively windows can be made to deal with summer temperatures that top 120°F. The palaces were mostly pavilions open to gardens through beautiful arches. Tapestries or silk banners were hung in the arches when privacy was wanted. Water splashing from fountains or down carved stone chutes inside the pavilions was channeled out into the gardens to thirsty trees and flowers, whose transpiration cooled the air. Windows were made from delicately pierced stone or wooden screens, whose tiny openings cooled the breeze as it came into the room.

Some pavilions were domed two-story structures with open cupolas on top. The sun would heat the top of the dome, causing the air within to rise out of the cupola windows, and draw cool air from the garden into the building below. The large interior space was also lit from above by light reflected off the roof into the cupola, and scattered by a thousand mirrors.

Time spent in such buildings is convincing proof that creative design for natural lighting, heating, cooling and ventilation has far more than just energy and economic benefits. Mechanically cooled buildings can be cooler, but they make us prisoners of their climate. Our metabolism doesn't adjust to the outside temperature, and we suffer from thermal shock every time we go outside. Cut off from the sights, smells and sounds of the outside world, our spirits suffer as well.

To join us to our world, windows take special forms. They bulge out to become bays. They settle down into window seats. They lengthen into French doors. They leap up onto the roof to become skylights and star windows, stand up higher to become clerestories, and push out from our houses to become gables and towers.

Builders of all cultures have crafted windows with special care to allow people to join in the passing parade — or just stand by and watch. The Nepalese sit in comfort on carved windowsills and chat with neighbors and friends in the

This window seat, in a castle in Turku, Finland, is large enough to be a room, but private enough for reading or prayer.

street below. Islamic women and children survey the dusty roads for signs of life through latticed windows, which permit them a view while keeping them hidden from the curious eyes of passers-by.

The owner/builders of California's redwood houses sometimes build window beds. In an alcove whose walls and roof are made of tempered glass, those in bed can look up at the stars twinkling through the pine needles or waken to the sun shining like diamonds through the raindrops tipping every leaf and twig.

To scatter the sun's light like a prism or soften it to a glow, craftsmen bevel and etch clear window glass. Stained glass artisans paint with light, creating pictures that fit within window frames. Sometimes these pictures incorporate the view outside through areas of clear glass; at other times, they hide it completely. It is up to the homeowner whether to accent the view or let the beauty of the glass itself capture people's attention.

The handcrafting of doors and windows can be merely a pretentious expression of wealth. Or it can be a quiet way of putting love and energy into a house so the house can in turn evoke love in others.

A well-crafted door or window performs all its functions with grace. It shows its maker's understanding of how such in-between places connect us to each other and to our world.

The craft that goes into a well-made door or window is not a luxury item. It is a productive investment that grows out of a different understanding of time and value than the "quick and dirty" items standard in today's construction. A

door or window carefully constructed with quality materials may cost a little more, but it lasts a lot longer. It works well instead of barely working. It gives us much more than the minimums of air, light and access required by building codes. It gives the satisfaction of good work to its maker, and delight to its users.

Tom Bender

THE MAKING OF DOORS

DOOR STRUCTURE

Some of our distant ancestors built fires on the thresholds of their caves to stave off the chill of night and the golden eyes glittering in the dark. Others improvised with skins and slabs of rock. The materials at hand and the level of technology achieved by a culture have always helped to determine how people enter and leave their shelters.

In the fertile Euphrates River valley, the first civilized peoples lashed together river reeds to form a door light enough that it could be lifted up and moved either to allow or to block entrance into a dwelling. The next innovation was probably the use of mud bricks and a wooden beam to frame a rough opening. Supplying hardware for the door required only a little ingenuity and the time to whittle wood or work hide into leather. This invented door is a remote architectural antecedent of the batten door still used today.

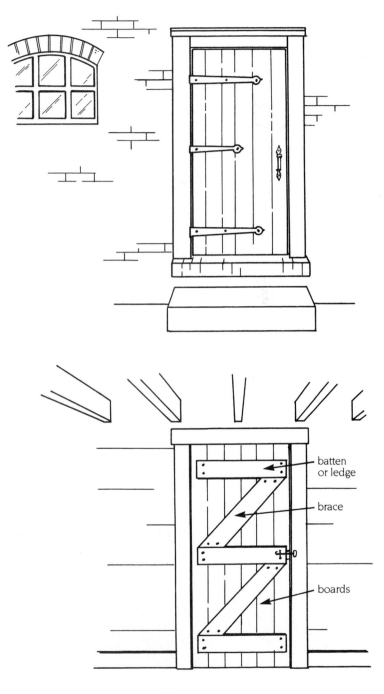

An exterior view of a batten door, top, and a batten door seen from the room side, bottom.

batten
or ledge

brace

boards

A batten door consists of a number of narrow vertical boards nailed snugly together against two or more horizontal crosspieces. The vertical boards may be square edged or tongue-and-groove. The horizontal crosspieces are called battens or ledges, and it is into the battens that the strap or T-hinges are screwed. Since the ends of the vertical boards are subjected to a lot of wear and tear when the door is in use, the top and bottom battens span them no more than a few inches from their ends, giving them strong support.

The sturdiest batten doors have at least three battens and two diagonal struts, called braces, connecting them. These braces are nailed between the battens from the free-swinging side of the door down toward the hinges. In this position, they strengthen the boards evenly and take some of the stress of supporting the door off the hinges. The braces may be butted against the battens or the battens may be notched to receive them. Both methods of joining the boards provide adequate structural strength for the door.

Both battens and braces may be chamfered or beveled along the edges to provide visual interest in what is otherwise a plain door. If battens and braces are on the exterior side of the door, they may be grooved with a drip kerf that runs the length of the boards along their undersides to encourage rain to run off the door rather than soaking into the boards.

The early colonists in this country occasionally studded a batten door with nails to make it look more impressive. Batten doors of the same period in England were not as well

The batten door above opens into a barn, and the one below serves as the entry to a potter's studio.

regarded. "The fate of the battened door," says H. Forrester, writing in *The Timber-Framed Houses of Essex*, "has been that which has befallen most constructional methods under the stress of new architectural forms. In the fourteenth century it belonged to the finest buildings. Then, abandoned in important houses except for doors of no significance, it continued to be the normal one for ordinary dwellings. Ultimately it was considered too inferior for the front entrance even of workers' cottages, and suitable only for back doors and those to outhouses." Batten doors are still more highly regarded in the United States than in England, and innovative builders here sometimes use them as entry doors on adobe and cedar-shingled houses. The style works well in rustic settings.

Large double batten doors proved an appropriate means of guarding the entrances to the fortified city-states of Renaissance Italy, but farmers found them suitable for more peaceable purposes. Double batten doors on a barn could easily admit oxen yoked to a ploughshare so that they could be unyoked and fed under shelter in inclement weather. What was good for the barn was also good for the stable. Ultimately, many of the "horseless carriages" manufactured by Henry Ford in the early 1900s were sheltered in garages with double batten doors when not left out in plain public view for everyone to admire. Although the overhead garage door is more common than the double batten door on garages today, such doors are still popular on barns and outbuildings, and are often used as entry doors in California and the Southwest.

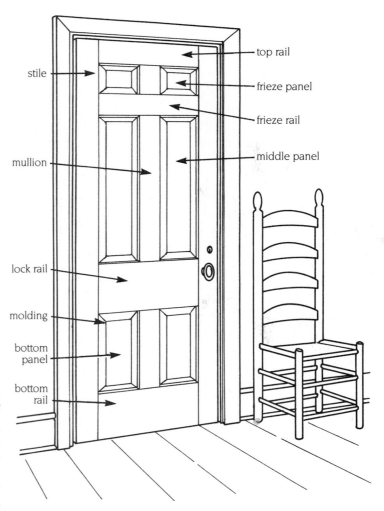

stile

top rail

frieze panel

frieze rail

mullion

middle panel

lock rail

molding

bottom panel

bottom rail

A panel door.

The panel door is not as old as the batten, but it has been more venerated during its history. The greater skill necessary in its construction initially made it a prestige item. The panel door of a nineteenth-century coal magnate's London house bore more resemblance to the columned temple portals of Greece or Rome than it did to the braced and ledged batten door on the house of his employee, a miner in Wales.

The coal magnate, however, could afford to hire the most skilled craftsmen to build his door. As a result, he enjoyed the advantages of the better protection from the weather that a panel door provides. Both sides of such doors could be made to look alike, and a greater variety of designs was possible.

The names of the craftsmen who "invented" the panel door are lost, but it's likely they were joiners doing a good business in batten doormaking, but frustrated by the problems of the batten door design. Because exposed end grain absorbs moisture more readily than long grain, the door's boards often warped; in addition, when cold winter air made the wood contract and the vertical boards shrink apart, the wind could whistle into the house through the gaps. To solve these problems, the joiners replaced the battens with a heavier framework of wood around the perimeter of the door, made into a rigid rectangle by glued and pegged mortise-and-tenon joints. The boards were fitted into a groove in this perimeter frame. No nails were needed, and a more finished-looking door resulted.

In later manifestations, the stiles, or vertical members of the framework, and the rails, or

Reading, Pennsylvania, once a booming railroad town, boasts many elegant panel doors such as these.

horizontal members, were fitted together with every possible variety of mortise-and-tenon joint, with dowels, and with other varieties of joinery, all in efforts to overcome the problems of an end-grain to long-grain joint. Additional vertical members (called center stiles or mullions) were often used to reduce the panel spaces between the stiles to the width of available boards and to lessen the amount of humidity-caused swelling and shrinking that had to be accommodated in each panel joint. Floating the panels in grooves in the edges of the frame allowed them to expand and contract in response to changes in the weather as well.

In Georgian England, raised paneling was common. Chamfering the stiles, mullions and rails was done for a decorative effect. Many varieties of molding were used to trim the panels, including the clamshell, scoop and quarter-round with which we are familiar today. The craftsmen of the time also created more elaborate styles of molding that are not duplicated by today's manufacturers.

The panel door lent itself well to imaginative design. The panels themselves could be rectangular, square, circular or teardrop-shaped to fit an arched doorway. Regency England's gentry used glass in fanlight transom windows to accent the beauty of their panel doors. Wealthy Victorians installed glass in the top or "frieze" panel to provide a window within the door itself. As the techniques of glass manufacture became efficient enough to support an industry, glazed doors became common.

With the introduction of industrial milling equipment in the mid-1800s, the process of making panel doors was both simplified and greatly speeded up. The mortiser, a machine that literally cuts square holes, now does in seconds work that might take a custom millworker or joiner of the last century several minutes to do with mortise chisel and mallet; the tenoner accomplishes a similar feat, making in one step what the home workshop owner must do in several even with power tools. Another sophisticated industrial machine, the single-head sash and door sticker, completely eliminates the need for making and attaching separate molding by cutting "profiled" molding out of the solid stile and rail lumber. The beveling of panels is done with a four-headed molder or shaper. With panel designs standardized, manufacturers save more time in the construction process.

The consistent quality of the materials used by manufacturers today gives them other advantages over the millworkers and joiners of the past. Our craftsmen ancestors tried to lessen strain on the joints of stiles and rails by making blind mortises, wedging their tenons, and using hide glues. Contemporary manufacturers solve the problem by working with lumber that has been kiln-dried to a relatively consistent and stable moisture content. In addition, they reinforce their simpler mortise-and-tenon joints with steel pins and flexible glues that can endure extremes of weather without hardening or breaking. Such glues are products of the technology that has emerged since World War II.

In the process of manufacturing panel doors, wise use is made of at least one of our natural resources — our forests. Guidelines established by the Architectural Woodwork Institute allow stiles, rails and panels of both exterior and interior doors to be made of a lower grade of lumber and faced with matching finish-grade veneers rather than being composed throughout of more expensive finish-grade lumber. This saves the manufacturer on lumber costs, and it lessens the demand for mature specimens of the rarer trees in our forest lands. The same interest in the bottom line has led some manufacturers who use applied moldings to cut these trim pieces from lower-grade wood and cover them with a plastic overlay — an economy that will not be greatly noticed once the door is painted.

The flush door, sometimes called the slab door, has been the chief competition for the manufactured wooden panel door since World War II. Plywood, glued over a perimeter frame or a core of solid lumber, gives flush doors excellent dimensional stability. These hollow-core or solid-core doors rarely warp or stick in door frames. They are simple and inexpensive to manufacture, and require less finish-quality wood in their construction than do panel doors.

The interior of a hollow-core flush door usually has very narrow wooden or cardboard stiffeners running vertically within the perimeter frame. The hinges are attached to this frame, and the lockset is supported by a solid wooden block positioned along the frame to house it. This lightweight construction offers no real protection against forced entry. A burglar can break through such a door with a well-aimed kick. Because it is little more than two sheets of plywood with an air space in between, a hollow-core door provides much less insulation value and fire protection than a solid wooden

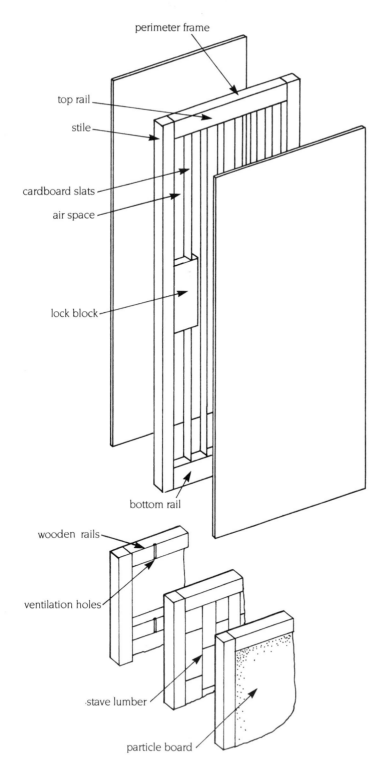

perimeter frame

top rail

stile

cardboard slats

air space

lock block

bottom rail

wooden rails

ventilation holes

stave lumber

particle board

Under its wood veneer face finish, a flush door may have a hollow or solid core within a perimeter frame. Cardboard slats (shown in the exploded illustration, top), or wooden rails (shown in the detail drawing, bottom), are used for the core construction of passageway doors. Stave lumber or particle board (also shown in the detail drawing) is used for the core construction of entry doors, as is rigid insulation.

door. It will not survive exposure to the changing seasons. For these reasons, a reputable builder will install hollow-core doors only on closets, bedrooms and interior passageways.

The solid core of a flush door may be made of particle board, fiberboard, rigid foam insulation or stave core lumber—blocks of scrap softwood machined to a uniform thickness and butted against one another on all sides, with a single vertical channel in the center to allow for expansion. A particle board or fiberboard core is rigid enough that it need not be reinforced with a perimeter frame, but a rigid foam insulation or stave core is usually edged with a narrow band of hardwood to improve the door's looks as well as strengthen it structurally. A solid-core flush door is sturdy and a relatively good insulator against both weather and noise. Although such a door is expensive, it is a good choice as an exterior door.

A solid-core flush door will usually have one layer of veneer glued to each side of the core, with the grain running at right angles to the grain direction of the core material. This "crossband" layer will be faced on both sides with a good veneer, called the face. A hollow-core door will usually have three layers of veneer glued to each side of the perimeter frame. These laminations strengthen the door and make it more resistant to warping.

If you buy a manufactured solid- or hollow-core flush door, it's most likely to be faced with lauan, a mahoganylike wood that grows prolifically in the Philippine Islands. Although lauan behaves like true mahogany, resisting the stresses of seasonal climatic changes well, it is more coarsely grained. Your building materials supplier may also sell flush doors faced with birch or oak. Finer hardwood veneers are available from custom doormakers—for a price. Cherry, mahogany, maple and walnut are just a few of the beautiful woods plain-sliced or rotary-cut into veneers for face-finishing custom-made flush doors.

In plain-slicing, the tree is stripped of its bark and cut in half lengthwise to establish a flitch. The flitch is sliced into long, thin sheets. These sheets are cut and numbered so that any irregularity in the wood will repeat itself as a pattern when the sections of veneer are butt-joined in sequence and book-matched. In rotary-cutting, the log, stripped of its bark, is turned on a huge lathe against a sharp blade, which cuts a continuous sheet of wood to a maximum thickness of $1/8$ inch. These sheets are also cut and numbered for book- or random-matching. A good custom doormaker will go to his lumber supplier and inspect the available veneers in the wood you have chosen to make sure they are of uniformly high quality and present a pleasing figure or grain pattern.

More exotic options for flush door faces can be found in reconstituted veneers. To produce a reconstituted veneer, a light-colored African hardwood such as ayous is plain-sliced into thin ($1/28$ to $1/32$ inch) sheets of veneer that are put into vats of colored dyes. After soaking for several hours, the sheets are dried and glued together in various color combinations to simulate expensive woods like Brazilian rosewood or American walnut. The reconstituted veneer is then cut into 2 by 11-foot sheets. Commonly

With a little imagination and woodworking skill, you can transform an inexpensive hollow-core flush door into something quite special. Craftsman Larry Golden used a saber saw to cut an opening in the lauan-faced hollow-core door above, then removed the cardboard slats that lined the door's interior so he could nail wood blocking around the inside of the opening to rough-frame a window. He used power tools to cut a sheet of Plexiglas to fit, lined the opening with stops, then put the Plexiglas in place. He finished framing the window by adding more stops and $3/8$-inch redwood trim. The contrasting wood tones and the unconventional shape of the window opening make what would otherwise be a very ordinary door into a real standout.

Reconstituted veneers, top, *add exotic good looks to custom-made flush doors, while plain hardwood veneers,* bottom, *show off the rich tones and textures of wood grain.*

used for fine furniture and paneling, reconstituted veneers make attractive and unusual faces for flush doors — but you will have to order from a custom doormaker.

The insulated steel door is the most recent innovation in the manufacture of doors, and it is one that the construction industry has been quick to adopt. The reasons are simple: An insulated steel door can be four times as energy-efficient as a solid wooden door, it is dishearteningly difficult for a would-be burglar to break through, and it effectively slows the spread of fire. A steel door requires little maintenance, and virtually no repair since it will not warp or split or rot as a wooden door sometimes will. Finally, it is usually designed to look like that old favorite — the panel door. Plastic moldings are used in place of wooden ones and insulated glass is the standard glazing material.

Manufacturers have taken considerable care to make sure the insulated steel door *works,*

usually selling it prehung in its own jamb, complete with weather stripping, a self-sealing door sweep, and a concealed drip kerf on the sill.

Like preassembled door systems for wooden panel and flush doors, steel door systems are easy to install. The unit must be fitted into the rough opening plumb and square; door trim is attached around the jambs to cover the joints. Hardware is installed. Other than this, there is not much to hanging a steel door.

However, there are some disadvantages to these doors. Although any steel door will support common cylinder locks, the narrow wooden interior frame is usually not wide enough to allow installation of a heavy mortise lock, which looks best on paneled styles. The surface will take paint well, but there is no way to make it take a stain. Steel can never imitate wood's variations in texture, grain and tone.

Building codes and manufacturers' standards have helped to upgrade the quality of entry doors all over the United States. Although there is no national building code, most townships and municipalities now advise building contractors to install some combination of storm or insulated doors and sash. Building codes also restrict the use of glass in and around a door, and dictate its kind. Although homeowners are usually free to vary design and placement of glass in entry doors, there are limitations on the number of square feet of glazing allowed. Check with your building inspector to find out the specifics of the codes that will apply to you, and you may save yourself some aggravation and expense.

All manufactured doors conform to standard dimensions. You can go to your lumberyard and buy a door 6 feet, 6 feet 6 inches, 6 feet 8 inches or 7 feet tall. The door may be as narrow as 1 foot 6 inches, but for an exterior door you would certainly want one 2 feet 8 inches or 3 feet wide. In some areas of the country, building codes require that each house have at least one entry door 3 feet wide to accommodate people in wheelchairs. The thickness of the door may vary from $1\frac{1}{4}$ to $1\frac{3}{4}$ inches. Interior doors are slightly thinner ($1\frac{3}{8}$ inches), and may be narrower as well.

Handcrafted doors can also be made to standard dimensions — or custom-fit for odd-sized openings. As long as certain elements of design and sound principles of construction are observed in making and hanging them, such doors will perform as well as manufactured doors, last just as long, and look even better. A batten door you make yourself can be rustic or refined, a panel door sturdy and plain or elegant, a flush door simple or exciting. When you make your own door, the choice is up to you.

Even though insulated steel doors imitate flush and panel doors in styling, they provide far better barriers to cold and forced entry.

MAKING YOUR OWN DOOR

The quality of the materials and the craftsmanship with which they are assembled underline and emphasize the special character of a handmade door. The tone and texture of the wood, the color of paint or stain used in the finish, the balance of proportions—of board to batten or stile to rail—can give the beholder a clue to the harmony of those within, their respect for tradition or the pleasure they take in departing from it. Whether you are a novice do-it-yourselfer or an experienced craftsman, you will discover that making your own front door affords you the opportunity to express yourself creatively.

Basic design considerations go beyond convenient placement of hinges, handles and locks. You must also take into account the setting for which you are making a door. A flush hollow-core door might be painted with a mural of a circus or a zoo if it opens into a child's room; the same kind of door, leading into a dining room furnished with Spanish-style tables, chairs and sideboard, might be dressed up with Mexican tiles framed with applied moldings. The architectural context in which a door will be seen can provide the inspiration for a striking and original design.

If you are planning to make an exterior door or two, there is an even larger context to consider—that of your house in its particular environment and climate. A jalousie door, with its adjustable louvered glazing and screen insert, might be perfectly appropriate for the sunporch entry of a Florida home; a Maine Yankee would feel more secure from the cold and wet of the long winter with a heavy, solid, wooden door that shuts into a weather-stripped door frame with an audible "thud."

The insulating value of your door depends on its design and the materials used to build it. Core construction and the seal around the perimeter are important. Where R-factor is the ability to resist heat loss, a hollow-core wooden door, with an R-factor of 2.18, comes in second to its twin in appearance, the solid-core wooden door (R-2.90). A wooden panel door (R-2.79) is almost as effective as a solid-core flush door. The energy-efficiency of all these doors is improved with the addition of a storm door, which creates an insulating pocket of warm air between itself and the entry door. A storm door made of aluminum, which conducts cold well, is less of a thermal barrier than an old-fashioned screen and storm door made of wood.

Glazing changes the energy-efficiency of a door. In general, the greater the area of glass, the lower the R-factor of the entire door. However, if your front entry is on the north side of the house and the interior is dimly lit, you may be willing

to pay higher heating bills to have more natural light. A single layer of glazing has an R-factor of 1; a ½-inch layer of double glazing has an R-factor of 1.61.

You must establish the practical as well as the aesthetic values you wish your door to serve—and be prepared to compromise some of these values to design a door that best meets your needs. Are you willing to trade a bit of energy-efficiency for light? Is natural ventilation important enough to you that you will repeat the seasonal rituals of installing and removing screen and storm door inserts? Will you feel safe behind a screen door or is security so important that you'll close your steel door and pay the heavy cost of central air conditioning? Energy-efficiency, light, ventilation, fire protection and security are all factors that must be taken into account in designing your door.

Ease of maintenance is another. A wooden door you make yourself will need as much upkeep and repair as a manufactured wooden door, and more than a manufactured steel door. In making your own wooden door, however, you have the opportunity to design and build to suit your needs, your climate and your location in city or town or country. You may take steps a manufacturer does not to insure structural soundness.

Material choices allow you to build and finish your door in ways that will lessen maintenance. Use kiln-dried wood to lessen the likelihood of the door's warping. Choose your wood for color, durability and weather resistance. Keep in mind that availability will affect cost.

Oak makes a strong and lasting panel door. Almost all the old-growth oak in the United States has been logged out, and the wood from younger stands, though expensive, is widely regarded as being of lesser quality. Walnut, another enduring hardwood, is even more difficult to obtain and is very costly as a result. Ash and birch too make good panel doors. All of these hardwoods need protection from the weather. A coat of paint will serve this purpose, but will hide your expensive and beautiful wood as well. Varnish or polyurethane will

give a clear, hard finish and show off that wood.

Redwood and cedar, both softwoods, weather beautifully, in time turning a soft silver. They resist rot well, and hold their shape after seasoning. Redwood, often used for panel doors on the West Coast, scars very easily and should be protected if the original finish is to be preserved. Cedar, which is less expensive, makes a fine batten entry door for cabins and other dwellings where the use of natural materials — fieldstone, local lumber and adobe, for example — is considered important.

Blonde woods like fir and pine darken and silver as they age if left exposed to the elements. They are relatively easy and inexpensive for the craftsman to come by. Although fir chisels and cuts well, it does splinter. Both fir and pine doors should be finished to seal and preserve the wood.

If you choose to make a flush door, a few softwood and hardwood veneer plywoods will probably be available to you at the local lumberyard or building supply. Fine hardwood and reconstituted veneers can be specially ordered. All plywoods have good dimensional stability because of their laminated structure, are relatively inexpensive because the core laminations are of lower-grade wood, and can show an attractive wood grain if care is taken in selection.

Hand or power tools can be used successfully to cut and assemble batten, panel and flush doors, but if you use power tools, take some safety precautions. Industrial woodworkers average twice as many accidents as workers in other industries in spite of management efforts to provide safe working conditions. Given that frightening statistic, the home workshop owner is foolish if he does not use the safest possible woodworking techniques, keep his tools sharp and in good repair and his workshop area clean and well ordered. A little sawdust left to accumulate on the workshop floor becomes a hazard in several ways: It may hide a power tool cord that trips you; it may start a fire if a spark should fly; it loads the air with particles that can irritate your lungs and eyes. Good house-keeping can eliminate this health hazard, and many others.

When working with power tools, remove jewelry, neckties and vests. Roll up long sleeves or wear a short-sleeved shirt. If your hair is long, pull it back out of the way and confine it with a hat or scarf. Wear a mask over your nose and mouth so you don't inhale tiny particles of wood, glass or sand. If you are going to use a power tool to cut, drill or sand, wear safety goggles. When the noise level in the workshop is high, wear ear protection. More than one wood-worker has gone partially or totally deaf because of continued exposure to the shrill whine of power woodworking equipment.

Make sure you know how to operate both power and hand tools safely. With portable power tools, clamp the wood you are working on to a bench or table so that it will remain fixed in position while you work. When using a table saw, brace the stock against the fence while ripping, and against the miter gauge while crosscutting. With a radial arm saw, brace the stock against the fence, clamping it if necessary, before drawing the saw blade through the stock. Pay attention to where your hands are and know

Woodworking Glues

White or polyvinyl glue, yellow or aliphatic resin glue, and plastic resin (also called urea formaldehyde) glue are good choices for use on interior doors, which are not subjected to weather extremes. When working with white or yellow glue, joints should be assembled within 5 to 10 minutes of application, clamped for 30 to 60 minutes, and allowed to set for 24 hours. Yellow glue holds with greater strength than white glue and stands up better to the heat of friction created by sanding, so it is preferred over white glue by many experienced woodworkers.

Plastic resin glue is sold as a concentrate that must be mixed with water before application. It can be used effectively only at 70°F or higher. It has long assembly and clamping times and, when dry, is brittle in joints that are not smooth and tight fitting.

Phenol-resorcinol glue, usually simply called resorcinol, is often used on entry doors because it is completely waterproof. It is a two-part (powdered catalyst and liquid resin) glue that must be mixed at a temperature of 70°F or higher. It has long assembly and clamping times, it bonds well, fills in gaps well and is impervious to fungus and temperature extremes. It leaves a visible glue line at joints, however, staining the wood dark purple-brown.

Generally, you should apply glue to both the wood surfaces you are joining. Some glues should only be used in a well-ventilated area. Read and follow the manufacturer's instructions for application, assembly and clamping to make effective and safe use of the glue you've chosen.

where the blade is, *even* if you can't see it.

Finally, know how you are going to accomplish your woodworking. If you know which machines you are going to use to do each operation and the order in which you will proceed, you can work with minimal distraction, concentrating on doing well whatever task is immediately before you. To accomplish this preplanning, work from a sketch of your design with all measurements marked.

The batten door is a good project for someone lacking woodworking experience. This simple door, often found on barns and other outbuildings, is also used as an entry door on cabins and adobe and cedar-shingled houses. A perfectly serviceable batten door can be made entirely of common board lumber, such as 1×6s, and a couple of handfuls of 6d or 8d galvanized nails. Tongue-and-groove fir and pine may also be used.

Calculate the number of vertical boards needed by measuring the width of the opening and dividing that figure by the actual (rather than nominal) width of a single board. You will undoubtedly have to rip at least one board to make up a door of the proper width. If you use tongue-and-groove stock, you'll want to rip the tongue off one board and the groove off another

to get a door with square vertical edges. Cut the boards an extra couple of inches long to allow you to square up the top and bottom edges of the completed door before hanging.

Use two sawhorses or a worktable to support the boards while you are assembling the door. Fit tongue-and-groove boards together or butt plain boards edge to edge. Clamp the boards with bar or pipe clamps to hold them in place while you nail the battens. Insert wood scraps between the board edges and the clamp's jaws to protect the door edges.

The simplest batten door is completed from this point by nailing two or three battens to the door boards. The easiest way is to nail through the battens and the boards, then clinch the nails; that is, bend over the ends that project through the wood. Doing this will strengthen the assembly—the nails won't be able to work loose—but it will mar the appearance of what will probably be the door's face.

It's more likely that you'll want the nails clinched onto the battens, which will be on the back of the door. To do this, you have to drive the nails through the boards. And to do that, you have to tack the battens to your clamped-together door boards, then turn the whole works over, supporting the ends of the battens with

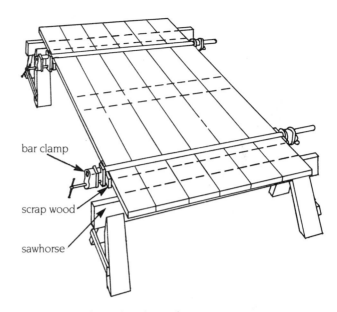

bar clamp

scrap wood

sawhorse

Boards clamped together for nailing through boards into tacked-on battens.

sawhorses. You have to support the battens or the nails will knock them off the door. And you don't want to nail the assembly to the floor or a workbench, so you've got to support it at least an inch or so above any such surface.

Braces, if desired, can next be cut to fit and nailed in place.

If the door is to be an exterior door, either use galvanized nails or be prepared to accept the stains that will accompany the rusting of ordinary steel nails. If you plan to clinch the nails, choose a size that's a half inch or more longer than the combined thickness of door boards and battens.

More sophistication can be obtained by using underlayment nails of an appropriate length. These nails have ringed shanks and will hold better than smooth-shanked nails. No clinching is necessary. Or use screws or carriage bolts or stove bolts with T-nuts.

A heavy, seemingly more substantial door can be constructed using dimension lumber, 2-inch and 3-inch material sold primarily for framing buildings. Or you can combine dimension lumber with tongue-and-groove lumber. Or use rough-sawn material obtained from a sawmill.

The hard edges can be removed by beveling, chamfering or rounding-over the edges of battens and braces. This is a particularly good idea on a door that will have its back turned to the weather. If you have a table saw or a router, plow a drip kerf along the bottom edge of battens; this will encourage water to collect there and drip off, rather than collect in the seam between batten and door board.

The strongest batten door has at least

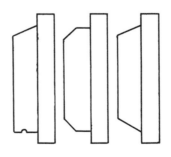

Three ways to encourage water runoff on doors with exterior battens—left to right, drip kerf, chamfers, bevels.

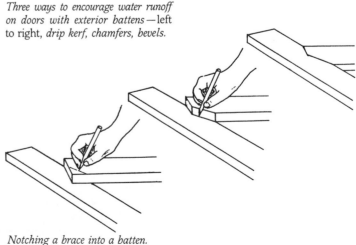

Notching a brace into a batten.

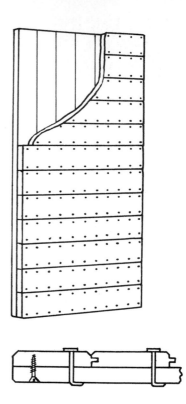

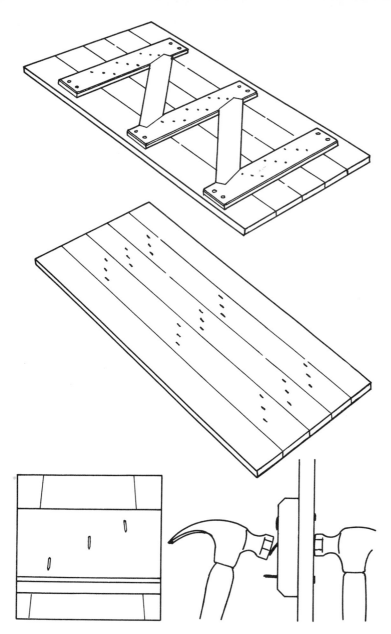

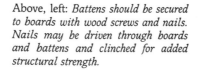

Above, left: *Battens should be secured to boards with wood screws and nails. Nails may be driven through boards and battens and clinched for added structural strength.*

Above, right: *Running battens to the full height of the door and studding with nails in a regular pattern makes a heavy, solid wooden door with Colonial good looks.*

three battens and two braces, with the battens notched to accept the ends of the braces. The braces should be positioned so they fall from the free-swinging edge of the door to the hinged edge. Obviously, you've got to know which way the door will swing and where the hinges will be before you construct such a door.

Cut your door boards to the appropriate length — somewhat longer than the doorway is high — and clamp them together. Cut your battens to length — slightly less than the full width of the door if there's any chance they would interfere with stops on the doorjamb — and lay them in position on the door. Usually, the top and bottom battens are 5 inches from the top and bottom edges of the door respectively, with third and fourth battens centered in the area in between. Tack the battens in place.

Rough-cut the braces, and lay them in place for final layout. Mark on both the battens and the braces where their edges touch.

If you want to install the braces *without* having them set into notches in the battens, simply scribe a cutting line across the brace's face from one edge mark to the other. Cut along the line, and the brace should drop snugly into place with its butt ends tight against the battens.

If you want to notch the battens, the layout is a bit different. If you've done the rough-cutting properly, one corner at each end of the brace should just touch the edge of the batten when the brace is properly aligned. If you haven't got this situation, mark the brace and cut it so you do. Now mark the centerpoint of each butt end of the brace. Scribe a line from that point to the mark you made to denote where the brace edge contacts the batten edge. Cut along this line. The brace is now ready.

Lay it in place on the battens. Mark the brace-end shape on the battens. Pry up the battens, cut the notches, tack the battens back in place, tack the braces in place, and fasten the door together.

The last step in constructing any door, and a batten door is no different here, is to carefully confirm the dimensions of the finished door-way, and check whether or not it is square. Then cut the completed but unfinished door to fit, allowing for the appropriate clearances.

The details of applying finishes to doors are in the following chapter, but you should, at this time, give some attention to the exposed end-grain wood in your batten door, especially if it is to be an exterior door. The end grain will absorb moisture quite readily and could begin to rot within a few years. Paint it with a wood preservative or seal it with paint or polyurethane

Joining the boards of a batten door with special flair, top, *makes the door very attractive. Varying the direction of the boards,* center, *creates an effect like that of the true panel door,* bottom.

or some other substance that's compatible with the finish the door will ultimately have.

Use strap or T-hinges extending from the edges of the door over the battens or over the areas of the face underlain by the battens to hang your door.

There are a number of ways to change the batten door for the better—varying its appearance and strengthening it structurally. Alternating two different widths of board—say 3 and 5 inches—makes the door seem more formal, as does a coat of paint. Running two layers of battens at 90- or 45-degree angles makes a heavy, solid door, not prone to warpage and capable of giving good insulation value. A Dutch door is a batten door cut in two across its width, and battened and braced in each part. A modified batten door with glazing can be made by framing the boards like a panel (just as the glass is treated as a panel) with grooved stiles and rails joined with half-lap joints, dowels and glue.

The panel door, being more sophisticated in its construction, requires more in the way of tools and skills. A manufacturer's woodshop will have heavy-duty industrial equipment to cut mortises and tenons, to bevel raised panels, and to profile moldings out of the stiles and rails. A reasonably skilled home woodworker can produce the same product with a few basic shop tools—like a table saw and a jointer—and key hand tools. Obviously, the more experienced you

are and the more caring you are, the better the door you produce will be.

Using a table saw, you'll be able to cut boards to proper lengths, rip boards to proper widths, make tenons, plow grooves, bevel raised panels, and cut a variety of moldings. With a good drill, electric or manual, you'll be able to start mortises and drill holes for locks, latches and other hardware. There are custom doormakers—true craftsmen—who make panel doors for production and on commission using just these tools.

Even power tools are not absolutely essential, although consummate skill then does become a requisite. You *can* make a panel door with only hand tools: handsaw, backsaw, drill or brace and bit, doweling jig, planes—including some fairly exotic ones—chisel, hammer, lots of clamps, try square and miter box. But if you haven't experience and skill with using the tools, your door may not look and function too well. It won't be possible to make moldings using these tools, but such a wide selection of moldings is available at most building supply stores that there should be no problem in finding a pleasing shape. Using these tools, you should be able to construct a door with mortise-and-tenon joints.

When planning your panel door design, keep in mind the usual relationships of the parts to one another. Top rail and frieze rail are usually the same width as the stiles and mullions (center stiles). The lock rail and the bottom rail are twice the width of the top rail. The lock is set 42 inches from the bottom of the door, level with the middle of the lock rail. Panels, in cross section, generally taper to a tenon approximately one-third the thickness of the stiles or rails, the same thickness as tenons used to join the framing members of the door.

Although the dimensions of the parts of panel doors vary, one custom doormaker interviewed for this book builds his 32- by 80-inch doors with these standard dimensions: Stiles measure 1¾ by 5 by 80 inches; top and frieze rails 1¾ by 5 by 22 inches, plus 2½-inch tenons on each end; lock and bottom rails 1¾ by 10 by 22 inches, plus 2½-inch tenons on each end. Panels vary in width and height according to their number and placement. However, they are likely to be ½ to 1¼ inches thick, depending upon whether they are flat or raised on one or both sides. If mullions are used to divide the paneled areas, their height will vary with the distance from top rail to lock rail and lock rail to bottom rail, allowing for tenons on both ends. Their width will be 4½ or 5 inches and thickness 1¾ inches, like the stiles.

If you want to make a similar door and you judge your skills to be up to the job, your work begins with the purchase of materials. You should have a plan drawn on paper and a materials list prepared. The list should itemize all the pieces necessary to assemble the door. Find a supplier who sells kiln-dried hardwoods and discuss your needs with him. The wood you want is expensive, sold by the board foot rather than the linear foot, and stocked rough-sawn in random widths and lengths. Within certain limits, you'll be able to pick and choose the particular boards you buy. You should be able to have them planed to a specific thickness for an additional charge, and unless you have an expensive thickness planer, you should pay for this service. If you do not have a jointer, you may also want to have one edge surfaced; the remaining rough edge will be removed as you rip the boards to the specific widths you need for rails or stiles or panels. Bear in mind that you should probably glue-up narrow widths to get 8- to 10-inch-wide boards for panels or lock and bottom rails, rather than trying to buy single boards that wide.

When you get the material home, give it some time to adjust to the temperature and humidity in which the door will exist. Pile the boards with 1-inch-square scraps of wood between them, so air can circulate completely around each piece.

Making the door itself starts with the layout of the individual parts. Crosscut each piece to length, then rip it to the desired width. Cut the stiles slightly longer than the height of the finished door, so you can square up the top and bottom of the door more accurately when the door is assembled and ready to be hung. When cutting mullions and rails, allow 2 extra inches for each tenon on the mullions and half the width of the stiles — 2½ inches where stiles are 5 inches wide — for each tenon on the rails. Lightly label each piece for identification.

Then cut the tenons. As a rule of thumb,

the tenon should be roughly one-third the thickness of the rail or mullion. It is best to make each tenon fit the mortise that will receive it as closely as possible (allowing for a fraction of an inch expansion or contraction with changes in the humidity from season to season). Mortising chisels cut recesses that are nominally ⅜, ½ or ⅝ inch wide; mortising attachments for a drill press cut square holes ¼, ⅜ or ½ inch wide. Establish the depth of the cut you make to remove waste wood on each side of the tenon by knowing the thickness of the cutting tool you will use to make your mortises.

To cut the tenons, remove the blade from the table saw and replace it with a dado head (with as many chippers as you have between the two blades) to make a wide cut. Adjust the height of the saw blade. Using the miter gauge to guide the cut, establish the shoulder of the tenon with the first pass over the dado head. Then make repeated cuts with the dado head — still using the miter gauge — to remove all the waste wood to the full length of the tenon. Turn the stock over and repeat the process. If the cut is slightly uneven, pare the sawn surface smooth with a chisel to finish-dress the tenon (without making the tenon too thin).

On certain of the rails, particularly the top and bottom rails, you need to create a shoulder on the edge. You cut this the same way you cut the side shoulder and cheeks, but the depth of cut of the dado head will have to be adjusted. If the depth of cut required exceeds the capacity of your dado head, you'll have to use a saw.

Make two tenons at each end of the lock rail and the bottom rail, but position the lower tenons on the bottom rail at least 1½ inches above the bottom edge. Otherwise, splitting may occur when the tenons are wedged during assembly of the door. To remove the wood between the double tenons on lock and bottom rails, use a band saw, saber saw or coping saw.

Next, plow grooves in the lower edge of the top rail, both upper and lower edges of frieze and lock rails, the upper edge of the bottom rail, both edges of any mullions, and the inside edge of each stile. The grooves may be cut with a dado head on the table saw, and should be ⅜ to ½ inch wide by ⅜ inch deep. The grooves will receive the panels.

Measure carefully to mark the position for mortises on stiles and rails. Use a drill to do the rough excavation, choosing a bit with a diameter matching the thickness of the mortise, if possible. Overlap the holes so you minimize the cleanup that must be done with a chisel. Use a doweling jig to guide the positioning and alignment of the holes. Cut 2-inch-deep mortises in the rails for the mullion tenons and 2½-inch-deep mortises in the stiles for the rail tenons.

As you proceed to clean up the mortise, fit particular mortise-and-tenon combinations. Lay out the framework and mark the parts so you know which tenon is going to go into which mortise. Then as you complete the mortises with the chisel, you'll be custom-fitting each joint. Try to make the mortises as smooth-walled as possible without giving up a snug, stiff, press fit.

When doing the fitting, keep the seasonal humidity variations in mind. As a general rule, fit the joints quite tightly during the more humid summer months and slightly more loosely during the drier winter months.

The panels for your door may be cut from

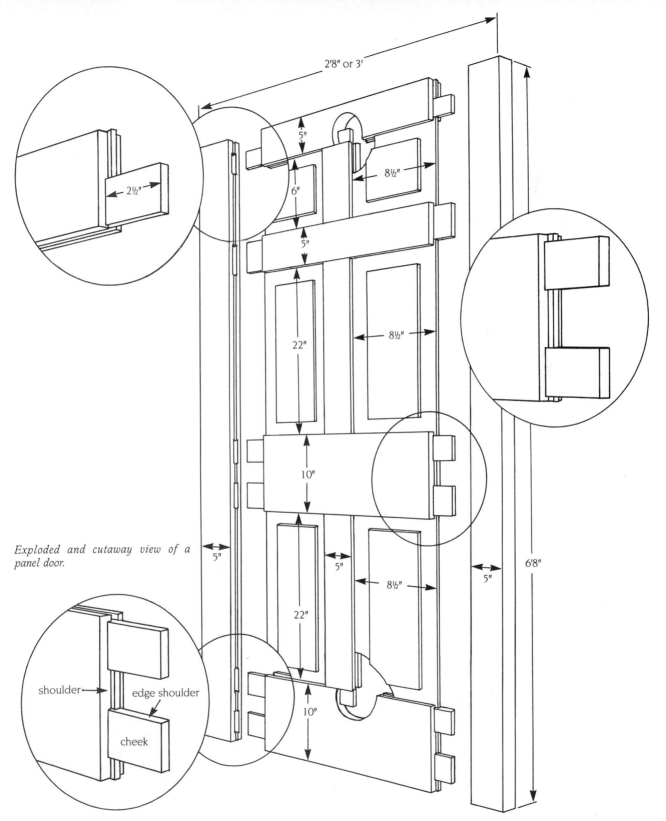

2'8" or 3'

2½"

5"

6"

8½"

5"

22"

8½"

10"

Exploded and cutaway view of a panel door.

5"

5"

5"

8½"

22"

6'8"

shoulder

edge shoulder

cheek

10"

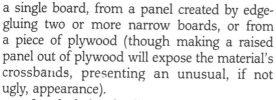

Painted or plain glass can take the place of a panel in a panel door for special effects.

a single board, from a panel created by edge-gluing two or more narrow boards, or from a piece of plywood (though making a raised panel out of plywood will expose the material's crossbands, presenting an unusual, if not ugly, appearance).

In calculating the dimensions of each panel, remember to add the depth of the grooves that will hold the panel to its length and width. Remember, too, that the panel shouldn't ride against the grooves' bottoms.

After the panels are cut to size and labeled for identification, calculate how to cut the kind of bevel you want. Determine the width of the bevel and the degree of taper. Decide if you want a lip between the bevel and the raised area of the panel.

Create the lip, if there is to be one, first. Set the depth of cut of your table saw's blade to the height of the lip. Set the fence as far from the blade as the bevel is to be wide (remembering to add the depth of a single groove to the bevel width). Make the cuts, doing both sides of the board if both sides are to have raised panels.

Cut the bevels themselves next. Position the fence so that the blade, when tilted, angles away from it. Then attach a high facing—perhaps 12 inches tall—to the side of the fence

closest to the blade. Set the blade to exact vertical and position the fence the correct distance from the blade for the edge of the bevel cut. Set the depth of cut to the width of the bevel, then tilt the blade in accordance with the taper you want. The cuts are made by standing the panel on edge and sliding it along the fence.

Be particularly careful when kerfing the boards for the lip and when cutting the bevels, since you'll have to remove the blade guard for these operations.

It is best if you rehearse the final assembly process. You want to be sure everything fits together as planned; not just individual joints, but the entire door. You also want to plan what clamps you must use and how they'll be positioned. And once the glue is applied, you want to proceed with a minimum of indecision and wasted motion. So practice.

It is a good idea to sand the individual pieces before assembly.

You may want to wedge the tenons during assembly, although not every craftsman will agree that this practice is necessary. Some believe that a snug fit, a proper spread of a good glue, and prompt and effective clamping will yield a strong and durable joint. But if you choose to try wedging the joints, make a cut into the butt end of the tenon, parallel to the height and almost as deep as the tenon is long. The wedge should be cut so it fits easily into the kerf initially, but spreads the tenon slightly as it's driven home. During final assembly, you must seat the wedges, spread the glue, fit the tenon into the mortise, and drive the pieces together. As you do so, the wedge will be driven home, spreading the tenon and wedging it solidly in the mortise. Of course, you can't use the wedges during your rehearsals, since you won't be able to get the pieces apart.

The glue you use should be dictated in part by the ultimate location of the door. An exterior door should be glued together with a waterproof glue, such as resorcinol glue. An interior door can be glued together with resorcinol or with yellow or white glue or whatever glue you normally favor.

As you assemble the door, spread glue thinly and evenly over the cheeks and shoulders of the tenons. Do *not* glue the panels in place; they must be allowed to expand and contract with changes in humidity, and gluing them in place could cause them to split or force joints open.

Spread glue and fit parts together in accordance with your plan, developed through practice. Install the clamps, repeatedly using a framing square and straightedge to affirm that the door is square and true. Tightening clamps can twist or skew the assembly, so you must check and recheck.

Put at least one bar or pipe clamp across the door at each rail. Extra pressure can be applied to mortise-and-tenon joints by applying a handscrew or C-clamp to each such joint to squeeze the sides of the mortise against the cheeks of the tenon. Scraps of wood should be placed between the metal clamp jaws and the door's wood to protect the door.

You should leave the clamps on the door

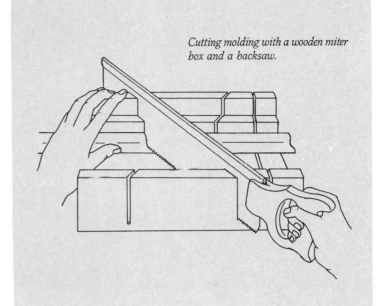

Cutting molding with a wooden miter box and a backsaw.

Miter Joints

A miter joint is simply a joint whose members meet at an angle — in most cases, 45 degrees. Over time, woodworkers have developed special tools to make cutting accurate miter joints a relatively simple task — wooden, reinforced, and metal miter boxes.

A wooden miter box is open on top and at both ends to allow you to place the stock you want to cut between the two box sides. These sides are slotted to hold a saw blade at 45 or 90 degrees to the wood. With the stock held firmly against one side of the box, you can use any fine-toothed saw to make your angled cut. However, a backsaw, a long, wide rectangular blade reinforced with a steel spine, is best designed to do the job.

The reinforced miter box has metal guides, as well as slots, to hold the saw in place for accurate cuts. A metal miter box is more expensive and more sophisticated, allowing you to cut many angles precisely.

For use in joining the moldings around the panels of a door, or the casings around the door frame, however, a simple wooden miter box will serve you well.

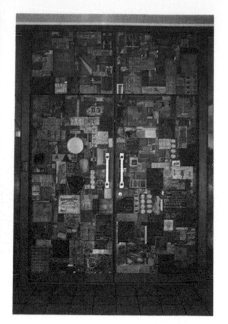

for about 24 hours, depending upon the glue. Read the directions on the glue container.

Only a few details remain to be completed after the door is unclamped. If you have left a little extra length on the stiles, cut it off so that the stiles end flush with top and bottom rail and the corners are square. If you plan to finish the door before attaching any hardware, sand it smooth, making a special effort to remove any glue lines visible at the joints. Then, even if you intend to paint or stain the door, seal top and bottom with polyurethane to prevent the door from absorbing excess moisture along these edges. (If you are an inexperienced woodworker, you may feel more comfortable finishing the door after you have attached hinges to door and jamb and adjusted the door to fit. As you become confident that you can make doors and door frames square and true, you will probably find it more convenient to finish the door before you frame and hang it.)

With the door assembled, you can add the molding that emphasizes the formality of the design. Attractive molding designs are available from any building materials supplier, but you can make your own using standard accessory bits for the router or a molding cutter in a table saw. Use a miter box to cut the corners of the molding to fit snugly against one another. Use brads and glue to attach the molding pieces to the frame. (Remember that the panels must be allowed to "float," expanding and contracting in response to the weather.)

This is but one way to make a panel door. There are others. Not only may other power or hand tools be used to vary the procedure, but numerous joints—both simpler and more complex—may be used to connect the parts of the frame. Glass may take the place of one or more panels to frame a pleasant view and add the beauty of light to the dramatic impact of the panel door.

Flush doors are available from manufacturers at such a reasonable cost that they are scarcely ever custom-made. They are, in fact, a product of post-World War II technology, and ideal for production with industrial equipment. A manufacturer can turn out flush doors so rapidly on the assembly line using such low-cost materials that the owner/builder is hard pressed to produce the same product investing no more in materials or in the equivalent dollar value of his own time and labor than he would spend to buy the door ready-made.

However, to say that flush doors are hardly ever handmade is not to say that they cannot be.

In fact, the same materials that save the manufacturer money and time spent in complicated joinery are readily available to the owner/builder at the local building supply. With plywood and particle board or fiberboard, you can assemble a flush door with a minimum of tools and little labor. Carefully consider the task before undertaking it, however. Will it teach you carpentry skills useful in larger projects? Will it acquaint you with tools with which you wish to be more familiar? If the final product looks rougher than the one a manufacturer delivers, will you be disappointed? If your reason for making a flush door is to provide yourself with a wooden canvas for painting or collage, would you do just as well to purchase a flush door and treat the surface? Or will part of your creative satisfaction derive from "stretching the canvas" yourself; that is to say, from knowing you have made the door from start to finish?

You cannot buy a manufactured batten door, so if that is what you want or need, undertake the project yourself. You can buy manufactured panel doors, but if you are willing to select materials with care and to learn to use power and hand woodworking tools with some skill and patience, you can improve on the quality, the craftsmanship, the durability and the style of what can be bought ready-made. A manufactured flush door, however, represents so efficient a marriage of technology and natural resources, that you should compare real costs with the dollar value of your time before beginning to build one.

If you decide you want to make a flush door, you will find a hollow-core door for interior use easy to build. For a door 30 by 80 inches,

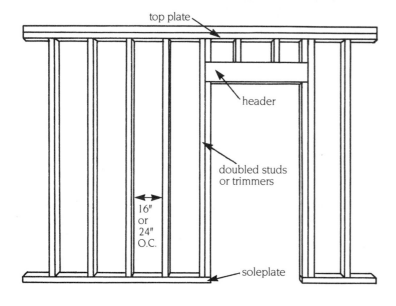

top plate

header

doubled studs
or trimmers

16"
or
24"
O.C.

soleplate

Rough frame for a door.

construct a perimeter frame of 1 × 6 stock. Cut two 80-inch pieces for the stiles and four 19-inch lengths for the rails.

Glue two rails in between the stiles at top and bottom. Position the other two rails within the perimeter frame equidistant from the end rails. Glue these in place and clamp the frame using pipe or bar clamps. Drive three corrugated fasteners at each joint, then remove the clamps.

Face the door with sheets of ¼- or ⅜-inch plywood. Cut two 30 by 80-inch pieces so that all you need to do is glue them to the frame with white or yellow glue. With either glue, you will have about ten minutes in which to straighten the plywood so its grain runs vertically and its edges are square. You may want to work with a helper.

You should clamp the newly glued assembly. At the least, attach handscrews or C-clamps at regular intervals around the perimeter of the door. If you have enough C-clamps and you can locate a couple of crooked boards that are as long as the door is wide, you can put some pressure on the central area. Select the edge of each board that has the crown and rest it on the plywood over those hidden internal rails. Clamp the board in place; you need not squeeze the ends of the board to the door to get the appropriate pressure at the middle.

You may choose to glue one face to

the door framework one day and the other the next day.

Drive ¾-inch brads around the perimeter, spaced at regular intervals. Set and putty the brad heads. Check the edges of the door for squareness and trueness. Plane to correct, then sand smooth. Apply a good finish to protect the surface and to discourage moisture absorption that might cause warping.

Making a solid-core flush door is even simpler using the manufactured materials available at a building supply store. Order one sheet of ¾-inch particle board and two sheets of ⅜-inch plywood cut to the finish dimensions of your door — in the example we have been using, 30 by 80 inches. Use resorcinol glue to bond the plywood to the particle board. Sand the edges smooth. Seal the edges of the door with polyurethane for protection from the ravages of the weather, and finish the door with paint or stain and top coat of your choice.

Framing a door begins with the framing of the rough opening for the doorway.

The norm is for the rough opening to be reinforced with additional 2 × 4s cut to the height and width of the opening. The header consists of doubled boards set on edge to span the width of the opening; doubled studs, sometimes called trimmers, run vertically between the sole plate and the header. The rough opening should be 2½ inches higher and wider than the door itself to allow for door frame construction and clearance.

Check the rough opening to make sure the header is level and the studs plumb. Check the subflooring at the foot of the rough opening to make sure both sides of the opening are level with one another. If they are not, you will have to trim one of the doorjambs to compensate.

If you are framing an exterior door, you must install a sill with a separate or integral threshold as a part of the door frame. Usually, exterior doors swing in to the interior of the house (except for screen and storm doors, which must open to the outside). As a result, rainwater, melting ice, snow and mudsplash can enter with the next friendly visitor, given a free ride on the bottom edge of the door. By sloping the exterior sill toward the outside of the house, you can do much to prevent this. If you use a storm or screen door, you will have to bevel the bottom edge of the secondary door to accommodate this slope.

If the threshold is an integral part of the sill, it should be slightly narrower than the door's thickness, so that the door extends over it, forming a natural drip kerf. Just underneath the front edge of the sill, another drip kerf should be cut to encourage runoff. On the interior side, the integral threshold should be beveled and rabbeted to form a protective lip over the finish flooring.

Most sills, including those produced by manufacturers and stocked at lumberyards, are made of oak. For centuries, carpenters have used this tough hardwood for sills, confident that it can take the abuse of countless feet and the repeated opening and shutting of doors. Modern technology does not offer a better alternative.

To figure out how to place your sill, determine the thickness of your finish flooring. Carpet is generally laid directly on the ⅝-inch plywood

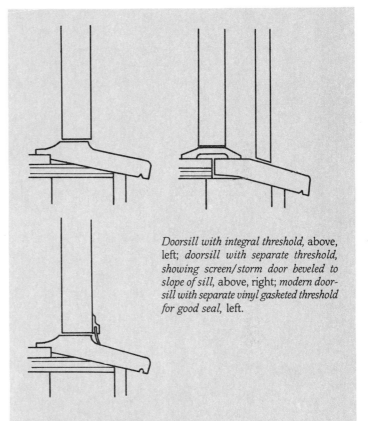

Doorsill with integral threshold, above, left; *doorsill with separate threshold, showing screen/storm door beveled to slope of sill,* above, right; *modern doorsill with separate vinyl gasketed threshold for good seal,* left.

An Energy-Efficient Door

The energy-efficiency of your door will be greatly enhanced if you make the seal between the door and the door frame tight. Use a manufactured door threshold, door sweep (a foam or vinyl flap attached to the front bottom edge of the door to seal against the threshold), or door shoe (a metal track fitted with a vinyl gasket that fits over the bottom edge of the door to seal against the threshold) to cut down on air infiltration and moisture leakage under the door. Seal cracks between the door and side and head jambs of the frame with vinyl or metal weather stripping (both of which are made with a spring flange to give a good edge and top seal) or with adhesive-backed foam weather stripping. Fit a separate wedge of adhesive-backed foam weather stripping against the strike plate. You'll find all these items, with instructions for installation, at a well-stocked hardware store or a building materials supplier.

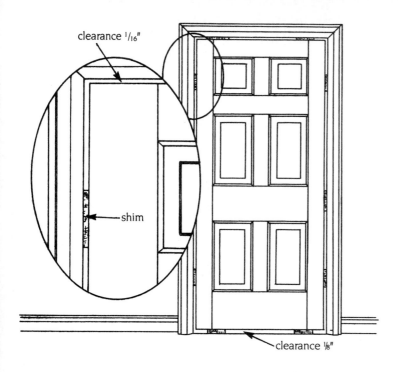

clearance ¹⁄₁₆″

shim

clearance ⅛″

Shimming a door square and true with wooden shingles or wedges before hanging for proper hinge placement.

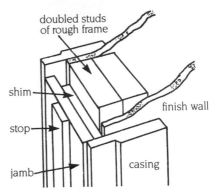

doubled studs of rough frame

shim

stop

jamb

finish wall

casing

Cross section of a door opening.

subflooring, as is hardwood flooring. Linoleum is laid on ¼-inch Masonite nailed to subflooring. The sill and threshold should rise ¾ inch above the finish flooring so that the door will clear carpet or throw rugs. To achieve this clearance, you will probably find it necessary to notch out the subflooring and supporting joists, but you may be able to nail the sill and threshold directly to the subflooring.

Interior doors do not need a sill, but often have a threshold to mark the change from one room to another and to bridge differences in finish floor heights. In interior door frame construction, the jambs are anchored to the doubled studs and the threshold is cut to fit between them.

If you were to purchase a ready-made door and frame assembly, you would probably find that the jambs were rabbeted or dadoed top and bottom to accept the head jamb and the sill. The doorstop might also be an integral part of the side and head jambs, formed by milling the wood. You don't have to fabricate your frame in this way, but you may want to be guided by what the manufacturers do.

Make the side and head jambs of ⁵⁄₄ lumber ripped to span the width of the doubled studs plus the finish wall surfaces. Cut the jambs square ended and long enough to span the height of the rough opening. Cut dadoes on each end of the jambs to receive the ends of the head jamb and the sill, if there is one. Cut the sill and head jamb to fit, keeping the dimensions of the door in mind as you work.

Fit head jambs and side jambs (and sill) together and position within the rough opening.

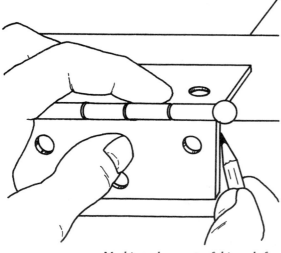

Marking placement of hinge before mortising.

Carving the surface of a flush or panel door can show the craftsman's sense of humor—or drama.

Check squareness. Wedge scrap wood—pieces of wooden shingles are often used—at intervals between the studs and the jambs to make the jambs absolutely vertical. A head jamb that is not level can be shimmed or planed to raise or lower one end.

Once you are satisfied with the squareness of the opening, nail through the jambs, through the shims and into the studs. Nail at regular intervals along the length of the side jambs. Nail on an angle through the dadoes to secure the side jambs to the head jamb.

Now attach the interior and exterior casing to the side jambs, and then to the head jamb. Nail the casing into the studs and jambs so that it covers the joint of the door frame with the siding, Sheetrock or drywall but leaves ⅛ to ⅜ inch of the jamb edge visible. This lip is called a setback or reveal. Sink pairs of casing or finishing nails at regular intervals along the casing. Then set the nail heads and putty before finishing the wood surface.

Hanging the door should be a relatively straightforward task if both the opening and the door are plumb and square. Make four wood shims equal in thickness to the clearances you want at top and sides (¹⁄₁₆ inch) and bottom (⅛ inch). The clearances at top and sides allow

for seasonal expansion; the larger clearance at bottom allows the door to clear carpet or rugs. Use the shims to wedge the door so that it will be positioned in the frame just where you want it to hang once it is on the hinges.

Measure for the placement of the hinges, remembering that hinges are attached to door and jamb on the side of the door toward which it will swing. An exterior door will be adequately supported by three 4-inch hinges, an interior door by two 3- or 3½-inch hinges.

Spread the leaves of the top hinge across

the doorjamb and door 7 inches from the top of the door and mark the placement with a pencil; repeat 11 inches up from the bottom of the door. Placement of the hinges 7 and 11 inches from the top and bottom of a door is traditional, but if you have made a panel door to a special pattern, you will want to make sure that the hinges will be set into the stile below the top rail and above the bottom rail so that you do not weaken your mortise-and-tenon joints. If a third hinge is to be used, mark its placement centered between the other two or, with a panel door, level with the middle of the lock rail (opposite the intended location of the lock).

After making these markings, remove the door from the opening. Clamp it or brace it so that the hinge edge is upright and level. Lay a hinge in position and use a utility knife to cut around it. Score the area within the pattern at ¼-inch intervals. Chisel out the wood; the hinge must lie flush with the surface of the door edge. Test the fit as you work. If the hinge does not lie absolutely level, chisel away high spots or putty low spots. Follow the same procedure to make mortises for the second and third hinges.

Attach the hinges to the door with screws. Unclamp your door and fit it into the opening

again with the shims to recheck the locations of the marks on the doorjamb. If necessary, correct the marks, then remove the door.

Just as you cut hinge mortises on the door, cut them on the doorjambs. Make sure that the hinges will be flush with the doorjamb surface when installed. Assuming you are using hinges with removable pins, pull the pins, install the loose leaves on the jamb, then mount the door by rejoining the hinge leaves and dropping in the hinge pins. If you have been careful to check the accuracy of measurements throughout this process, the door should swing freely, with adequate clearance all around. If it does not, you may be able to adjust it to work well by loosening the screws in the hinge leaves on the jamb a turn or two.

Finally, attach the doorstop to the jamb, so that the door closes against it.

All that is necessary to finish the job is to install a lock assembly or doorknob and latch. Install a complementary strike plate on the jamb. Drill with a bit and auger or an electric drill to create an opening for the lock or latch. If a mortise is required in the jamb, excavate it with a chisel. Manufactured locksets come with detailed installation instructions.

Hanging It All Up

Remember the children's nursery rhyme that begins, "There was a crooked man who had a crooked house . . ."? The truth about the doors shown right is less whimsical, but more interesting. A Hollywood film executive commissioned an inventive, talented craftsman to build a set of doors for his new home, and crooked doors is what he got. Both doors are operable and will swing freely without sagging — causing anyone who appreciates the difficulty of hanging symmetrically designed double doors to remark with wonder.

Larry Golden, the craftsman who made the doors, visited the building site in California to get accurate measurements from the rough opening. Then he returned to his woodshop in Wisconsin, where he built both the doors and door frame. By preassembling the entire door unit, he was able to insure everything fit and functioned properly.

When Larry installed the door unit, he took the doors out of the frame he had built, cut off the jamb extensions to make the jambs fit the rough opening, and then toenailed the jambs into the sill. He wedged shims between rough openings and jambs to make the final opening plumb and true and rehung the door. This simple procedure is the same one used to install manufactured door units. The owner/builder can save himself time and aggravation by adopting it to hang his handmade doors.

FINISHING TOUCHES

Your door is assembled. It is joined with screws and nails or with intricate joints and glue. It is level and the corners are square. You have built this door to last, and you take pride in your workmanship. The finishing touches you choose will add even more to its enduring value.

You have many options. You can trim panels with molding or inlay bands. Decorate a flush door with marquetry. Paint a panel door to match exterior house trim. Use a flush door as a canvas for a mural in a child's bedroom. You can stain your door to accentuate beautiful grain, then apply a finish that will allow you to feel the texture of the wood or protect it with a hard surface coating.

Once the surface is finished, you must add hardware that will make the door perform the functions appropriate to its setting. Every entry and passageway door will need hinges, but not every door will need a lock. Doors in and out of a kitchen may not even need a doorknob. When made with a small window to watch for incoming traffic and hung on double-acting or spring hinges, a kitchen door fitted with simple pushplates and kickplates will allow the cook in the family to carry dishes hot from the oven right into the dining room without unnecessary juggling.

Hardware may be chosen with more than mere function in mind, however. The quality of materials and construction helps determine the life span of the hinges and locks on a door just as it determines the durability of the door itself. Style is also an important consideration: Forged iron strap hinges and thumblatch seem to belong on a batten door; a shiny, polished brass, mortise lockset and lion-head door knocker make a front entry door more impressive; handmade hardware on any door makes it something very special.

Don't be overwhelmed by all your options. It is the purpose of this chapter to help you explore and evaluate them, so that you can choose those that will add the most to the looks and life span of the door you have crafted.

Decorative wood trim, if any, should be added before you finish the surface of the door. Traditionally, batten doors are used for outbuildings and dwellings made of rough-hewn wood or stone, and are not trimmed. Flush doors, surfaced with lauan or more exotic woods, are usually also left undecorated. However, if you want to imitate the look of a panel door without undertaking its more elaborate construction, you can apply half-round decorative moldings to the flush door surface. If you have

A variety of simple and ornate moldings, such as those shown at left, can be purchased from building materials suppliers and mail-order companies to decorate panel or flush doors.

made your own panel door, you will want to buy or make moldings that will emphasize its beautiful symmetry, and you may even want to construct an elaborate architrave to make a formal entrance.

Building supply stores have both hand-carved and heat-pressed or embossed hardwood moldings in straight lengths, and mail-order woodworkers' supply companies sell rounded corner and arc sections. Half-round, quarter-round and bolection moldings (rabbeted out on the underside to fit over the joint of panel and stile or rail) are easy for the do-it-yourselfer to install.

After you have chosen the style of molding you want to use on your door, measure it and cut it to the correct lengths with a handsaw. Bead a water-resistant or waterproof glue along the edges of stiles, rails and mullions and press the molding in place around the panels. Use a hand drill to make starter holes for brads, and tack the molding in place at the corner joints and regularly along its length. Remember that the panels in your door must be free to "float," expanding and contracting with the weather. *Don't* nail your molding to the panels. Secure it by nailing on the diagonal into the stiles, rails and mullions.

Moldings are available commercially in mahogany, maple, oak, poplar and walnut. If none of these hardwoods can be stained to match the wood you have used to make stiles, rails, mullions and panels, you may want to make your own moldings.

Manufacturers use a sticking machine or molder to make moldings integral to the stiles, rails and mullions. Professional woodworkers fit their spindle shapers or routers with carbide or high-speed steel cutters, which are available in various patterns, to cut integral moldings.

An ordinary router can turn out beautiful molding. Steel and carbide bits are available to cut many different molding profiles. Whatever bit you choose, make sure that the shank will fit your router's collet.

If your workshop is reasonably well equipped, you'll probably have tools to make curved and carved moldings to accent rounded or teardrop or S-curved panel doors. Without a spindle shaper, cut curved molding with a band saw and bevel it with chisels. Use other carving tools to create more elaborate decorative effects.

If you are confident of your carving skills you may want to try your hand at incised or raised carving on the panels. You can get a similar effect by using wood ornaments, available through mail-order woodworkers' supply companies. These embossed wood carvings, which are made in standard medallion, shell and trim patterns, can be secured to the door surface with glue and finished with paint or stain to match.

Inlay and marquetry can also be used to dress up interior doors. If you have made a panel door, you can use inlay banding around the perimeter or within the panels to emphasize the door's strong lines. Simply rout out a groove, using a straight bit as deep and wide as the inlay banding. Cut lengths of the bands to fit your design, mitering the ends so the bands will join at the corners. Use contact cement or white or yellow glue to glue the bands into the groove, pressing them flush with the frame or panel

If you have a router, you can make your own moldings by using special bits. The ogee bit, top left, cuts an undulating curve, top right; the rounding-off bit, bottom left, cuts a stronger, rounded curve, bottom right.

A panel door intended for interior use can be "dressed up" with decorative inlay bands. After assembly but before finishing, rout out grooves to the depth of the inlay, apply glue, and press in the bands.

surface. Use a very fine grade of sandpaper to remove any glue lines. Cover the panel on both sides of the inlay banding with masking tape, then paint white shellac over the inlay banding before you use stain or filler on the door to prevent the inlay from absorbing any colored pigments. When the white shellac has dried, remove the masking tape and proceed with finishing.

Just as there are standard-pattern ornaments for those who are not skilled at carving, so there are inlay rounds, ovals, rectangles and corner designs for those who want the effect of marquetry. A flush door surfaced in an exotic veneer might be made more attractive by the careful placement of these decorative veneer pictures.

Inlay rounds and ovals showing flower, leaf, shell, eagle and more abstract designs come from the mail-order woodworkers' supply companies with a protective paper coating. All that is necessary to install them is to rout out an area of the door surface to the dimensions and shape of the inlay, and glue it in place. When the glue has dried, you can wet the paper coating and sand it away to reveal the design. Use a very fine grade of sandpaper and a light touch! If you plan to stain the rest of the door, protect the inlay round with white shellac, just as you would inlay banding. Such delicate effects are suitable only for interior passageway doors that will not be exposed to the weather.

Finishing the door is necessary to stabilize the wood's moisture content, whether you plan to hang it in the artificially maintained climate of your house or to expose it to the extreme effects

of sun, wind and rain. All construction-grade lumber is kiln-dried to a moisture content of 19 percent or less. Most craftsmen air-dry their lumber another three to six months in their woodshop or shed before they use it. Even wood seasoned this way expands and contracts with changes in humidity, however, so finishing is necessary to minimize moisture absorption.

Your surface finish will be protected from weather extremes and will require much less maintenance if you make your first handcrafted door to fit an interior passageway. To quote from the U.S. Department of Agriculture (USDA) publication, *The Wood Handbook,* "Good finishes used indoors should last much longer than paint coatings on exterior surfaces. Interior finishing differs from exterior chiefly in that interior woodwork usually requires much less protection against moisture but more exacting standards of appearance and cleanability."

If you choose to paint interior doors and trim with gloss enamel or semigloss paint, you should prepare your wood surfaces as carefully as you would for staining. They should be as free of imperfections as possible. If you plan to use a flat latex paint, your surface preparation need not be so elaborate. Be forewarned, however, that a flat paint on doors and trim will be easily dirtied. Even if the manufacturer describes it as "washable," it will not take a good scrubbing the way an enamel will.

Prepare your interior door for painting by smoothing the wood surface. Set nails visible at joints or in molding and fill with wood putty. Raise hammer marks and other dents by steaming the damaged wood area with a hot iron over a damp cloth. Fill any visible seams at joints with wood putty.

After the wood putty dries, begin sanding. File especially rough edges that show saw marks. Use a medium grade of sandpaper to smooth all the door's surfaces to a uniform cut. If you are hand-sanding, make sanding blocks (rectangular for flat surfaces, curved to fit moldings) to insure an even treatment. Wrap them with felt, and then with sandpaper, to prevent the sandpaper from "gripping" the wood. Work *with* the grain.

If you own a belt sander, you can use it for rough to medium sanding, but be sure you sand *with* the grain, and be careful that you don't accidentally gouge the surface or round over an edge. To get a truly smooth surface, final sanding should be done with a finish sander or by hand.

Use progressively finer grades of sandpaper to go over the whole door. If you begin sanding with 120-grit sandpaper, you may want to use 150, and then 220 to smooth the surface enough to take an enamel finish. If you intend to use a flat paint, a surface sanded to a "uniform cut" with 120-grit paper will be smooth enough.

To finish-sand a softwood surface, first wet it with a damp sponge to raise the loose fibers that feel whiskery. Allow the wood surface to dry, then sand with 220 paper. About half an hour later, dampen the wood, let dry, and sand again. This step not only finishes the preparation of softwoods before enameling, but is essential in readying any wood surface for a water-based stain.

Hardwoods that are open pored, such as ash, lauan, mahogany, oak and walnut, need to be sealed with a wood filler and lightly sanded again before being painted with enamel. Thin a paste wood filler to a creamy consistency with turpentine following the manufacturer's recommendations. Brush the filler on with the grain, then across the grain, making sure you completely cover the wood surface. The object is not to apply a uniform thickness of filler onto the wood surface, but to work what is applied deep into the wood pores.

After about half an hour, scrape off the excess filler with a putty knife, working across the grain. Be careful not to gouge. Use a coarse cloth to clean away any remaining excess, again working across the grain. Let the wood filler dry at least overnight before painting with enamels.

If you have made a door out of pine, you should seal any knots with a coat of shellac so that the resins in them will not "leak" into your enamel, altering its color. It is not a bad idea to shellac the entire surface of any white pine or ponderosa pine door for this reason.

Finally, the door's surface is ready to be painted with enamel. First, apply one or two coats of primer. Once the primer has dried, sand it lightly with fine sandpaper for a super-smooth surface. Wipe away any dust with a tac cloth, then paint on your finish coat of enamel.

If you are finishing a panel door, paint the panels first, working from stiles toward the mullions and from the top of the door to the bottom. Next, paint the top rail, frieze rail, lock rail and bottom rail with long, smooth strokes. Finally, paint the stiles. When one side of the door is thoroughly dry, paint the other side. Do not neglect the door edges, for the crosscut end grain of the stiles will absorb moisture more readily than any other part of the door unless properly sealed. In addition, the door edges take some abuse every time the door is opened and closed.

If you are enameling a flush door, paint in small overlapping sections, brushing each section vertically and then horizontally. Work quickly, so that overlapping paint marks will not show. Paint the door from top to bottom. Allow to dry. Paint the other side and the edges. The result of all your effort will be a beautiful door in the color of your choice with a smooth, hard finish, and a satiny or glossy sheen.

Painting entry doors does not require such meticulous surface preparation. Repair any blemishes in the wood and sand with 120-grit paper. Coat the door with a good water-repellent and preservative, such as Cuprinol or Pentothal, making sure to adequately cover all joints and end grain. When using such toxic chemicals, protect your eyes, your skin and your lungs. Allow at least two days for the water repellent/preservative to dry before painting.

Use an oil or alkyd-oil based primer if you want to paint your redwood or cedar door; use an acrylic latex primer for doors of pine or Douglas fir. This is the most effective way of sealing in the resinous substances that seep to the surface of these woods over time. For other species, you may use whatever primer you like so long as it is nonporous and free of zinc-oxide pigment, which tends to make the top coats blister.

Two layers of a good-quality trim paint over your primer will give you a finish that lasts

eight to ten years. If you choose an oil-based paint, don't allow more than a week between application of the first coat and the final coat. This will help to prevent peeling. Work during the middle of the day when the temperature of the door and the air are most stable to prevent the paint from blistering, wrinkling and losing gloss. Latex and alkyd paints are easier to work with than oil-based paints, but any good-quality trim paint will give you a durable finish that will be simple to maintain.

Finishing a door with paint is not your only option. If you selected your wood for the beauty of its grain, you may want to stain or varnish it. Before settling on a finish for it, however, you must decide what color stain, if any, you want to use on the wood, and how smooth or rough you want the texture to be.

Stains may be water, alcohol or oil based and contain varying amounts of pigment. Some contain so much colored pigment that they look like a thin coat of paint. The amount of pigment also determines the degree of transparency or opacity. Stains color wood either by penetrating into its pores or by building up a surface coating.

Finishes work in much the same way: penetrating into the wood grain or building up protective layers on the wood surface. Tung oil, which comes from the nut of the tung tree, sinks into the wood and repels surface moisture. The popular Danish oil finishes use tung or linseed oil as a base and phenolic resins or urethane as hardening agents. They penetrate into the wood so it retains its natural texture, and slightly harden the wood to protect it. Because of the way Danish oil finishes work,

Top: *The use of a clear finish (by Deft) and a variety of stains (by Minwax) makes this wood mosaic section of a sugar pine panel door glow with warm tones.*
Bottom: *Glid Tone Oil Wood Finishes in Golden Oak, Cherrywood, Honey Maple, Weathered Barn and Walnut,* left, top to bottom, *are comparable in color and coverage to Minwax's Golden Oak, Cherry, Colonial Maple, Driftwood and Special Walnut,* right, top to bottom, *when applied to scraps of pine.*

41

Homemade Stains

Charlie Southard, a custom builder in New Mexico, has developed a method of mixing stains that he uses to beautiful effect on both exterior and interior woodwork, as evidenced by the door above.

He uses Stoddard Solvent, available from wholesale petroleum product distributors, as a base. He experiments with artists' oil paints, sold in tubes at art supply stores, to add color. He finds that a ½- or 1-inch length of paint mixed into a quart of the base makes enough homemade stain to finish a door.

Charlie suggests substituting a mix of mineral spirits and linseed oil for the Stoddard Solvent if this product is not readily available. He offers one final tip: Keep a record of the formulas you use to save yourself some time and effort should you ever decide you want to finish another door or window to match.

they are also called penetrating resin finishes. There are few Danish oil finishes formulated by manufacturers for exterior use, but many of the craftsmen interviewed for this book use clear Deft or Watco Danish oil finishes to bring out the beauty of their entry doors. They apply multiple coats before installation, then renew the finish every year or two thereafter. The pure urethane or polyurethane finishes coat the wood surface and harden to a glasslike smoothness. You see the texture of the wood, but you can't feel it. These finishes are extremely durable and are frequently intended for exterior use.

Whatever combination of stain and finish you decide on, your interior or exterior door must first be sanded to a very fine cut if the treatment is to effectively show off the wood's beauty. You can prepare the surface just as you would for enameling, with a few qualifications: Whatever wood patching material you use to cover blemishes must be compatible with the stain you have chosen. Seek advice on whether or not it will "take" water- or oil-based stain. Remember, too, that if you apply multiple coats of a clear finish, the wood putty must match the color of the wood. If you cannot find a patching material of the right color, you can mix glue and sawdust from the wood to create your own putty. In any case your patch may absorb stain differently than the wood itself, so experiment on scrap lumber until you get the desired results.

Before final sanding, wet softwoods down to raise the grain, especially if they are to be finished with a water-based stain. Do not treat porous hardwoods with a wood filler until *after* the first coat of stain is applied, so that you can make a color match.

Before applying any stain, dust surfaces of the door with a tac cloth. Wipe off any grease or dirt with a cloth wetted with mineral spirits. How you apply the stain depends in part on what kind of stain you choose. Wiping stains are brushed on for complete coverage and allowed to sink into the wood before the excess is wiped off. Water stains are flooded onto the wood surface and allowed to dry. Non-grain raising stains, which are alcohol based, work like water stains, but are less forgiving of a clumsy application. Since end grain is more absorbent than long grain, it may stain darker. Prevent this by thinning the stain or removing the excess stain more quickly from end-grain areas than you do for the rest of the door. Whichever stain you choose, closely follow the manufacturer's instructions, and test your results first on scrap wood. Take it through the entire surface preparation and finishing process.

After you have stained and sanded and stained again and finally achieved the tone you want, apply the top coat. Your choices are an oil-based finish to which hardeners have been added or a pure urethane or polyurethane. Follow the manufacturer's instructions for easy application.

Hardware is installed when at last your door is finished. The range of style options available today is wide enough to suit houses built 200 years ago or just last week. The owner of an eighteenth-century restored city brownstone can use a mail-order catalog to find brass finial

hinges, rim lock, and nameplate knocker for his paneled entry door. A farmer determined to install sturdy strap hinges on his new double batten barn doors can probably pick them up at his local hardware store.

Door hardware can, of course, be handmade. Hinges and handles can be carved from wood, forged from metal, or cut from leather. Latches, slide bolts and barrel bolts can be cast in iron or chiseled out of oak. It is a rare craftsman, however, who can make a lock secure enough for use on an entry door. For this reason, those who want to use handmade hardware should install it only on interior passageway doors or exterior doors where security is not a problem.

Traditionally, certain styles of hardware are associated with certain styles of door. The

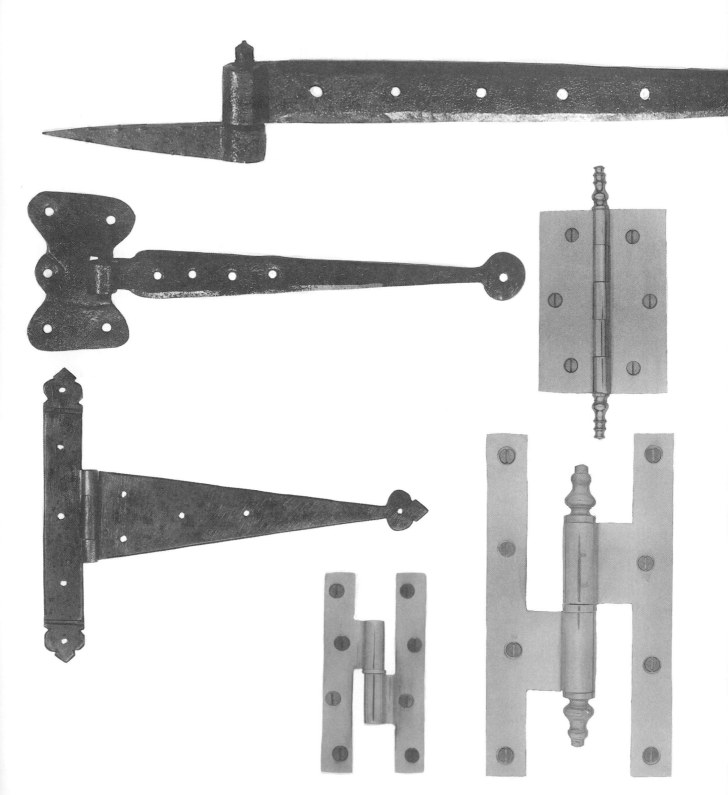

rustic batten door seems to demand the support of iron strap hinges. The hinges may be secured to the frame by a T- or butterfly-shaped hinge plate. Use of a "pintle," an upright pin welded to a nail or screwlike base, is less common.

Although strap hinges of forged or cast iron are somewhat rough looking, do not think that they are without decorative value. The long hinges end in an area of metal called a cusp that may be shaped like an arrowhead, heart or bean. Centuries of blacksmiths working at the forge have even established regional designs. The Moravians of Bethlehem, Pennsylvania, decorated hinges and latches with a tulip-shaped cusp; New Englanders inscribed their iron hardware with a Salem Cross.

Panel doors, with their elegant balance of lines and proportions, have traditionally been installed with butt hinges. However, doors in Tazewell Hall, a house built in Williamsburg, Virginia, about 1725, were supported with brass H and HL hinges. The hinge plates of an H-hinge are longer and narrower than those of a butt hinge; the HL-hinge combines the virtues of an H-hinge with those of a strap hinge by extending a metal "leg" several inches across the rail. H- and HL-hinges are generally used in combination, with HL-hinges at top and bottom rail and an H-hinge at lock rail level. Although they may not be available at your local hardware store, you can order handsome reproductions through mail-order companies such as Renovator's Supply, and Ball and Ball, listed in the appendices.

The standard butt hinge used today is a descendant of the Baldwin Patent hinge first

Opposite, far left: *Handforged or manufactured hardware can add to the style of your door. The forged iron hinges shown (by blacksmith Newton Millham) are especially appropriate for a batten door. The heavy-duty strap hinge, top, meant to be secured to the door jambs with a drive pintle, is intended for use on exterior doors. The lighter strap hinge, center, is meant for use on passageway doors. The cross garnet hinge, bottom, is more decorative and formal in feeling.*

Opposite, near left: *Solid brass hinges look well on elegant panel doors. The butt hinge, top, is not handed, but the knuckle hinges, bottom, are. The lack of a finial head on the hinge pin of the smallest knuckle hinge makes it seem more modern in feeling than the other two (all by Omnia Industries).*

Above: *A handcrafted batten door hangs on one-of-a-kind hinges.*

cast in iron around the time of the Civil War. This hinge was manufactured so that it could not be taken apart. This innovation made the hinge easier to install and more secure against forced entry.

Hinges today are usually made of rolled brass, although some manufacturers still cast them. Iron butt hinges are available from manufacturers of period hardware. Hinge pins are usually made of brass or plated steel, which resist binding. When purchasing hinges, look for cut (rather than sheared) edges on the hinge plates, square corners, and knuckles that meet precisely and move freely.

Two other varieties of butt hinges are also quite popular today on both panel and flush doors: the loose-pin hinge and the loose-joint hinge. The loose-pin hinge has three separate parts—the right-hand hinge plate, the left-hand hinge plate, and the pin. After hinge plates are attached to door and jamb with screws, the knuckles are aligned and the pin dropped in. A widened head on the pin stops it from dropping right through the knuckles to the floor, and allows the door to be removed easily from its frame by prying up and taking out the pin. Loose-pin hinges may be decorated with a ball or finial on the pin head to add to the hardware's formal good looks.

The loose-joint butt hinge is simpler in design than the loose-pin butt hinge, but functions in a similar way. Each of the two hinge plates has only one knuckle, but the knuckle on the hinge plate that attaches to the jamb has the pin welded securely into it. The knuckle of the other hinge plate, which attaches to the door, simply fits over the pin and pivots around it whenever the door swings. A door supported by

such hinges can be removed from its frame simply by lifting it upward, so that the knuckle on the hinge plate attached to the door clears the pin.

Other butt hinges that may be appropriate for use on your handmade door are the rising-butt hinge and the ball-bearing hinge. These are useful in special applications: The rising-butt hinge, with its angled knuckles, lifts a door's edge up to clear carpet as it swings open. The ball-bearing hinge, which turns on lubricated ball bearings instead of a pin, offers extra support for any heavy door.

Hinges may have two, three or as many as five knuckles for strength in supporting a door. At least two hinges, and usually three, should be used to fasten the door to the door frame. These hinges should be placed so that the weight of the door is evenly divided between them. Carpenters have a rule of thumb for placement—7 inches down from the top of the door and 11 inches up from the bottom, with the middle hinge centered between the other two. If you have made your panel door in a nonstandard size, align the bottom of the top hinge with the bottom of the top rail, and the bottom of the lower hinge with the top of the bottom rail; center the third hinge between them.

When installing your hinges, make sure that the hinge screws are long enough to go through the jambs and shims and into the doubled studs. Hinges attached only to the jambs can be pried loose by a determined prowler with a crowbar.

Since panel and flush doors are built symmetrically, hinges and locks may be placed on the left or right depending on how you want the door to open. Although by law the doors of all

public places must open outward for fire safety, the front doors of private homes generally open in. If your front door opens to the inside and toward the right, it is "right-handed"; if it opens in and to the left, it is "left-handed."

The hand of any door in a house can be determined by standing on the corridor side of it and noting the direction of its swing. Should a door open toward you, it is described as right- or left-hand reverse bevel. This term refers to the beveling of the edge that bears the lock.

Although standard butt hinges are not handed, loose-joint and rising-butt hinges are. In purchasing these, you must make sure that you obtain correctly handed hinges or you will not be able to install your door properly.

Interior passageway doors generally open in and closet doors out. However, there are some good reasons to make exceptions to the rule. Where two doors are at right angles to each other, varying the direction of the swing and the placement of hardware can prevent the doors from opening into one another.

In addition, it is important that door handles be placed on the same side of the door as the nearest light switch. No one who's had a long, hard day needs the added frustration of having to grope for the light switch. This may determine the hand not only of your entry door, but also your bedroom and bathroom doors.

In picking hardware, think seriously about who will use the door and how they will use it. You may not want a keyed lock on the bathroom door if a child uses it — or you may deliberately choose a privacy lock so simple in its design that you can pick it yourself with a bobby pin or a dime.

For a swinging door you want only a simple pushplate to protect the door where it gets the most abuse. The pushplate is the simplest of all door hardware. The door pull, which can be attached by itself to the door or mounted on a pushplate, serves the opposite function.

The next level of complexity in door handle hardware is the thumblatch, which combines a door pull with a movable thumbpiece that lifts a bar latch on the opposite side of the door. Thumblatch sets are available in cast, stamped or forged iron or cast or stamped brass for use on batten doors.

The term latch refers to any bar that will spring back into its keeper or strike plate when you close the door. In the case of a simple barn door latch, this happens because of gravity. In a cylindrical, a mortise, or a rim lock, the latch is spring-loaded so that it pops out as soon as you release the doorknob.

A close relative of the old-fashioned latch is the bolt, which may be horizontally or vertically mounted on the door. The surface bolt and the cane bolt are sturdier than a standard latch

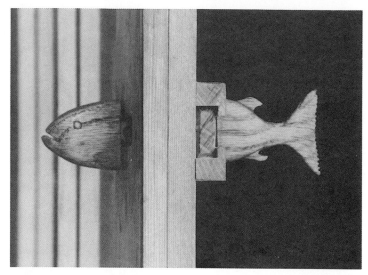

When you know how to use woodworking or metalworking tools, you can make handmade door hardware to suit your own tastes — fanciful, funny or sturdy and practical — whatever you choose.

not only because of their thickness, but also because bar guides and keeper are welded onto metal plates.

Most of us have never pushed back a surface bolt to open a barn door, but we are used to turning round, polished brass knobs to enter the house, the bedroom and the bath — not only in our own house, but in the houses of friends and neighbors. The popularity of the polished brass finish may fade, but cylindrical locksets are here to stay because they offer the builder the most convenience and ease of installation at the least expense.

In general, a cylindrical lockset consists of two doorknobs, a connecting spindle, and a spring-loaded latch housed in cylindrical casing. When the latch is thrown, it protrudes through an opening in the faceplate (attached to the door's edge). If the door is closed, the latch will jut into a matching opening on the strike (installed on the jamb). A decorative round metal plate, called a rose or rosette, is mounted with long screws behind each doorknob on the door's surface to hide the spindle hole.

Cylindrical locksets offer minimal to good security. Inside the house, use a passage lockset, which will latch but not lock, wherever only a minimum of privacy is required — on a closet or hallway door or a door leading into a den.

For greater privacy, you can purchase privacy locksets (rim and mortise as well as cylindrical styles). Intended for interior use, privacy locksets offer a turn button to lock the door from inside bedroom or bath and a knob so simply "keyed" that it can be picked from the corridor side with a bobby pin. This is a useful compro-

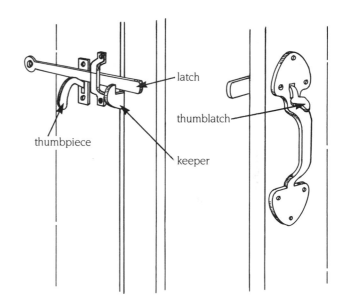

The parts of a door latch, as seen from both faces.

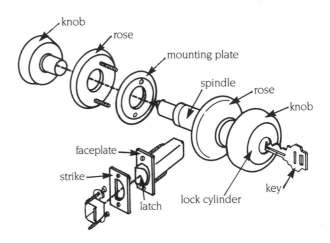

Exploded view of a circular lockset.

mise for anyone living with small children or elderly relatives. It provides them with a real sense of privacy, but allows you to come quickly to their aid in the case of fire or accident.

Cylindrical locksets designed for use on entry doors are keyed from the exterior side and may be locked by turning or pushing a button on the interior knob. Unlike a privacy or passage lockset, an entry lockset should have a deadlatch (a metal rod like a miniature deadbolt) behind the lip of the regular latch to help prevent forced entry. The quality of the materials used should be high so that the lockset is durable. As in other matters, you get what you pay for. A heavy-duty residential entry lockset may cost three times as much as the no-frills model usually installed by general contractors, but it is built to last and a better deterrent to burglars. Of course, a good lock by itself does not protect your entry.

Margaret Miner, in an article on home security in *The Old-House Journal* (November 1981), described the half-thought-out solution that satisfies most homeowners: "Millions of homes are outfitted with expensive locks attached with tiny screws to flimsy doors in rotten frames." If you have gone to the trouble of handcrafting your own entry door and frame, you have taken a first step in making your home less attractive to burglars. Carry through when you buy a sturdy entry lockset and hinges and attach them to your door and frame with *long* screws. If such details are important to you, check out other high-security door hardware options.

If you have a paneled entry door, a mortise lockset is a good style and security choice. The

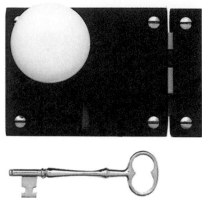

Lockset styles are available to suit many different architectural styles and many different tastes.

Above: *A rim lock and bit key, in black on brass and solid polished brass (by Baldwin Hardware Manufacturing Corporation).*

Right: *Three mortise locksets: "Lexington," in polished brass finish, with separate lock cylinder and tumbler, near right; "Richmond" in antique brass finish, center; and "St. Augustine," in satin brass finish, fitted with a left-handed lever, far right (by Baldwin).*

Opposite, left to right: *An interconnected entry and deadbolt lockset in polished brass finish (from the H series by Schlage Lock Company); a crystal doorknob; a cast brass doorknob, depicting the face of the North Wind; a solid brass lever, fluted in a style from the reign of Louis XV (all from Renovator's Supply); a rosewood doorknob (by Baldwin); a porcelain doorknob, painted with apple blossoms (from Renovator's Supply); a white porcelain lever (by Baldwin).*

mortise lockset gets its name from the location of the lock mechanism. All of the lock's moving parts are enclosed in a steel housing designed to fit in a mortise in the stile of the door. For this reason, a mortise lockset cannot be installed in a door with stiles less than 1⅜ inches thick and 3 inches wide.

Generally, the locking mechanism of a mortise lock is a spring-loaded, curved or beveled lip latch and a deadbolt. The latch may be three-fingered for more even wear and easier closing. The strike of a mortise lockset is handed to receive the curve or bevel of the latch. Both deadbolt and latch can be unlocked by turning the key in the deadbolt cylinder. The combination of a double lock and a protected locking mechanism is enough to give burglars real pause—and if they pause long enough, they'll either get caught or go somewhere else.

To purchase a mortise lockset, you must know the hand of the door and the backset dimension—the measurement from the edge of the door to the center of the lock. If you want to install a mortise lockset in a steel door, get one with a narrow backset—that is, if your steel door will support the lock. Most steel doors are hollow and the false stiles too narrow to contain the housing of a mortise lockset, although a few steel door manufacturers install lock blocks within the stiles to give their customers this option.

Owners of paneled entry doors on Colonial houses often opt for a rim lock, so called because the locking mechanism sits in a box on the interior rim or surface of the door. Two hundred years ago, Benjamin Franklin probably opened his front door with a bit key or a skeleton key—the same kind he used in his kite experiments. He lifted a keyhole cover, inserted the bit key *through* the entire thickness of the door into the rim lock, and turned. His front door then opened with a push.

Since his time, the appearance of rim locks has changed little, but manufacturers have improved the locking mechanism. Rim locks made for today's exterior doors may hide a deadbolt cylinder under the keyhole cover; the deadbolt is operated by a standard cut key. The bit key may be used from the inside of the house to lock or unlock the deadbolt as an alternative to a turnpiece—or may simply be hung by the door frame as decoration.

If you were to ask a policeman which piece of door hardware is the single most effective deterrent to breaking and entering, chances

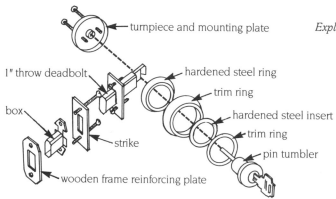

- turnpiece and mounting plate
- 1" throw deadbolt
- hardened steel ring
- trim ring
- hardened steel insert
- trim ring
- box
- strike
- pin tumbler
- wooden frame reinforcing plate

Exploded view of a deadbolt lock.

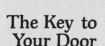

The Key to Your Door

If you have ever played "tug of war" with a key in a lock, you were probably using a key cut from a duplicate rather than from the original or were trying to work a cheap lockset—or both. Third- or fourth-generation duplicates will work, with a little force, in cheap locksets. But high-security locksets, which are manufactured with a minimum of variation from parts specifications, won't accept such imprecisely machined keys. Only the original key and accurate duplicates made from it or from the locksmith's code for the original key will turn smoothly in the lock cylinder.

Protect your home by building doors and door frames soundly, installing high-security locksets, and being very particular about whom you trust with duplicate keys. And if you move into an existing house or formerly inhabited apartment, have all the entry locks replaced or rekeyed. Landlords at "high-security" apartment buildings or complexes will perform this service as a matter of course at no charge to the tenant. If yours won't, pay for replacing or rekeying the locks yourself. It is the only way you can make sure that strangers of unknown character do not have keys to your door.

are he would answer, without any hesitation, the deadbolt.

A good deadbolt is made of solid steel or brass reinforced with steel dowels, and usually has a 1-inch throw. When thrown, a deadbolt cannot be pressed open with a credit card and is a challenge to cut with a hacksaw. A single-cylinder deadbolt installed on an entry door may be opened with a cut key from the exterior or by turning a turnpiece on the interior. There are variations that are operable from one side only by key or turnpiece. These are of value if you are concerned with protecting yourself and your family while you are in the house.

The double-cylinder deadbolt, used most often on doors with windows or sidelights, offers double protection. Not only must you use a key to get in, you must use a key to get out. This means that a burglar who comes in through the window while you're out won't be able to steal a television or stereo or piano—or anything he can't lift out that window.

If you go this route for your entry doors, you may want to get all your locks keyed alike, inside and out. In case of fire, you don't want your family to have to search for the right key. Some people make emergency exits easier by leaving keys in the keyholes of the interior deadbolt cylinders in the locked position whenever they're at home, and taking them along when they go out. Another option is to not use the inside deadbolt cylinder when you're at home—but then why pay for it? Because double-cylinder deadbolts kept locked from the inside can

slow escape from a fire, they are illegal in some states. Check with your local building inspector.

As is common in building, there are trade-offs between fire safety and security. Discuss the problem with your family. Make them aware of escape routes they can use in case of fire. Make sure there are enough keys to go around or that children know where to find the extra key should they need it. Let everyone know that keys should not be lent out to casual friends, and that any lost key may mean rekeying all the lock cylinders.

Like locksets, keys have become more sophisticated to improve security. The bit key of a couple of centuries ago was simply designed, with sometimes only one cut on its shank that had to align with the locking mechanism to release the lock. Most modern door keys have five to seven cuts on their shank and work a pin tumbler inside a cylinder. They are easily duplicated and are not terribly precise.

The most sophisticated key discovered in researching this book is made in Sweden and distributed in this country by locksmiths. The Kaba key fits a pin tumbler, but this pin tumbler has pins in each of three quadrants that must fall precisely into cuts along the top of the shank and both sides of the key to release the lock. To cut a Kaba key requires an extremely precise drill press that costs about $4,500—about ten

times the amount most locksmiths spend for equipment. As a result, the keys are not available in all parts of the United States. Check with your local locksmith or write LORI KABA, Old Turnpike Road, Southington, CT 06489.

A word about hardware styles: Manufacturers make their door hardware to standard sizes. As a result, you can sometimes mix and match to get the decorative effects you want. If you think your front door would look good with a satin brass handleset on the exterior and a round white porcelain knob on the interior, you can probably get the hardware you want to make it work.

The variety of styles seems almost endless, although the functions door hardware serves are narrow. A knocker or doorbell or buzzer lets you know when visitors arrive; a wide-angle viewer tells you who they are. Nameplate and house number help the mailman find you; a mail slot lets him drop mail into your house, out of bad weather.

Whatever hardware you choose, let it express both your intelligence and your taste. Like the finish on your door, the hardware has certain functions to perform. If it achieves or exceeds your expectations, you'll know that you have chosen well. You should take pride not only in your handcrafted door, but also in your finishing touches.

A GALLERY OF DOORS

Tom Barr

GREENHOUSE ENTRY DOOR

When Tom Barr and a few friends decided to start Mad Dog Design and Construction Company in northern Florida in 1973, they were intent on building and remodeling houses with the high standards and originality of craftsmen. "We saw ourselves as artists," Tom says. "We got into doing doors with stained glass, and spiral staircases. We had a good feeling for working with wood." And it was obvious that other people also felt they had a good feeling for working with wood, because their reputation spread.

Three years later, in 1976, the Bettons saw their work and asked the company to build a small deck and greenhouse addition for them. The front door, which echoes the latticework frame of the greenhouse in its design, was a special commission—a surprise birthday present to Mrs. Betton from her husband, built in secret and installed while she was away.

Tom designed and built the door. He used cypress in the panels and for the central frame laminate and red oak for the exterior and interior frame laminates. All the lumber for the door was logged in Florida, rough-milled, kiln-dried, and seasoned for several months.

Tom surfaced and trimmed the red oak for the frame laminates all at once. Since the frame has three layers, Tom concerned himself more with planing the inside edges of the frame pieces than with cutting them to exact dimensions. "It didn't hurt a bit to use boards of varying width for the stiles and top and bottom rails," he says, "and it eliminated a lot of the problems I might have had in trying to keep everything dimensioned just the same way. I didn't even try to trim up the door until it was all glued."

Tom used a table saw to cut mortises through the ends of stiles and top and bottom rails. Then he set his radial arm saw to 45 degrees and cut the ends on the diagonal (to miter the corners without the aid of a miter box). He repeated these steps on a smaller scale to make the frame for the large window sash.

At the center of the top and bottom rails, along the inside edge, Tom made a blind mortise to receive the tenons cut on the mullions. He prepared the upper and lower members of the center sash in the same way. He cut a third pair of blind mortises in the stiles at the level of the lock rail. Then he made tenons on both ends of the two lock rails and all mullions.

Tom assembled the first frame laminate by joining the mitered ends of stiles, top and bottom rails and sash with narrow splines and Weldwood plastic resin glue. When he was sure that the inside corners were square and all the joints tightly aligned, he clamped the frame assembly. He used glue blocks between the clamps and the frame to prevent the wood from marring, and let the glue dry for the recommended time. Then he put another frame laminate together in the same way.

When it had dried, Tom had identical door frames, with smooth and flat inside edges. He put the interior frame face down on his worktable, and reinforced each corner joint of stiles and rails with a corrugated metal fastener.

Next Tom attached the "spacer," the central frame laminate. He nailed and glued lengths of ¾-inch-thick oak against the fasteners to create a narrower perimeter frame on top of the frame already on his worktable. He also butt-joined lengths of the ¾-inch oak down the mullions and around the sash to finish off the support structures for the panel lumber. (If you decide to build a door like the one Tom made, you need not use ¾-inch lumber. What is important is that you make the central frame laminate the same thickness as the panel lumber so that all the joints come together properly.)

Tom prepared the cypress for use as paneling by sanding it with 120-grit paper on an orbital sander, and jointing the edges. With a radial arm saw set at 45 degrees, he sawed the ends of the cypress boards so that he could lay them inside the framework in a pleasing diagonal pattern. After eyeing up the first board and cutting it properly to fit one of the corners, he checked the length of the longest side of it with a tape measure. This gave him the measurement

Above, left: *The wooden structural supports and window mullions make a pleasing latticework in the Bettons' greenhouse entry.*

Above, right, and right: *The buttresses of the arch, laminated of several thicknesses of board glued together, are pegged to the rest of the frame for structural reinforcement and visual interest.*

for the short side of the next board; the longest side of the second board gave him the measurement for the short side of the third board, and so on. In this way, he was able to cut his panel lumber without a pattern.

Tom glued every other cypress board against the frame to allow the panels some room to expand and contract with changes in humidity. Then he stapled all of the boards in place with 1¼-inch staples — long enough to penetrate through the boards into the frame without protruding on the other side. He completed the assembly by driving corrugated metal fasteners into the corners of the exterior frame, on the inside face. After applying more glue all along the top of the spacer, he flipped the exterior frame over so that the fasteners were against it. He squared up the edges, secured them with clamps, and hammered the exterior frame onto the spacer, completing the triple-laminated framework of the door.

After the glue had dried thoroughly, he removed the clamps and trimmed the outside edges of the door to rough size. Then he planed these edges by hand to their finish dimensions.

To make the two small windows in the door, Tom cut enough red oak to trim both inside and outside, mitered the corners, and joined them with glue. Then he lined interior and exterior window trim up snug against the stiles, and screwed right through trim and panels to the trim on the opposite side. With a saber saw, he cut out the paneling inside the window trim.

Tom used a router with a ⅜-inch-radius ogee bit to round all the inside edges of the oak door frames and window trim. Then he sanded the red oak with 200-grit paper and oiled the entire door with Watco floor finish.

With the finished door flat on his worktable, interior face down, he applied stops to the interior side of the openings. He fitted a single layer of glass into each of the openings, tacked in glazing points, and puttied.

Tom hung the door in its frame with three sturdy 4-inch butt hinges. To open and lock the door he added a door handle handcarved of cherry and an ordinary deadbolt. He shaped the interior handle with a round dowel that fits through the door, and the exterior handle with a round stem mortised to receive it. He split the end of the dowel, started a little wedge in it and brushed on Weldwood plastic resin glue. Then he joined the parts of the handle together inside the mortise cut for them in the door.

When Tom visited the Bettons to arrange photographs for this article, he was not surprised to discover that the door was covered with mildew. He rented a high-pressure steam cleaner to clear away the unsightly mold, and then treated the door with a product called CWF, which is both a sealer and a fungicide.

Experience with mildew problems has taught everyone at Mad Dog that finishing wood in a humid subtropical climate requires special care. "Lately we've been using Thompkins Watersealer," Tom says, "and we mix a fungicide right in with it. It's a very effective product."

An innovative company, Mad Dog experiments not only with new products, but also with management styles. It is no longer run as a woodshop co-op as it was in the early days, but it is still very democratic. It has grown to a full-time crew of 48, and each week all its members meet to discuss project responsibilities and progress, to participate in planning sessions, and to continue to define the company's new goal—to build affordable passive solar housing for people like themselves. "Mad Dog is really a family business, even though nobody's related," Tom says.

Now that the consensus of this "family" is to design and build inexpensive, energy-efficient shelter, they subcontract the custom work they used to do to local craftsmen. This means that Tom isn't making any more doors like the one he made for the Bettons. He's glad for the woodworking experience he gained doing such custom work, however. In a company determined to make a positive impact on the housing industry in northern Florida, it is good to have people on hand who can speak from the perspective of the craftsman.

Bob Jepperson

HANDMADE DOOR HARDWARE

Most of Bob Jepperson's bicycle rides are over. In 1981, he and his wife built a small hydroelectric power system to bring power into his house and woodshop, so he doesn't have to go three miles each way to a friend's barn to share the use of a table saw, a band saw and a drill press.

However, you can still find him walking his own back acres in Washington State looking for lumber. What he likes best is to find a fallen fir tree that has been lying undisturbed on the woods' floor, seasoning for half a century. The bark may be crumbling and the sapwood soft, but the heartwood of an old-growth tree is wonderful woodworking material. Since such finds are rare, Bob uses other sources as well. He looks for fruitwoods such as apple or pear in his travels around the countryside, and yew and hardwoods in the state forest (the wood is sold by the state to private citizens for $20 per

1,000 board feet). He asks friends who are loggers to keep an eye out for special wood to add to his stockpile. Then, after the chill of winter has set in and the trees are dormant, he cuts down the ones he has selected and brings them back to his woodshop for milling.

To speed the air-drying, Bob often cuts the logs down into 1-inch planks. Then he brushes the crosscut ends of prized boards with old paint to prevent them from drying too rapidly and checking. He dabs melted paraffin on the end grain of precious woods for the same reason. The lumber is stored in a barn for at least a year, then brought inside the shop for another year of seasoning. Once the moisture content has stabilized, he surfaces the planks and laminates them with glue to build them up to whatever thickness he needs. "Air-dried wood can be just as well seasoned as kiln-dried," Bob says. "In all the Stradivarius violins, there wasn't a single piece of kiln-dried wood. Proper seasoning by air-drying is just a matter of time."

Bob enjoys the sculptural quality of door hinges and handles and has a file full of designs he'd like to try. However, when he's commissioned to make a door with handmade hardware, he does his best to find out what the client likes and dislikes. He sketches his ideas on paper until he comes up with one he feels he can get really excited about, and then takes it to the client for approval.

For strength, Bob likes to make the top and

Left, top: *For a rustic effect, Bob chisels the wood roughly and leaves counterbored screws exposed.*
Left, center: *This hinge is curved to fit a batten door on a root cellar.*
Left, bottom: *The curly maple hinge, also shown opposite, is used at the level of the lock rail on this plain panel door.*

61

bottom hinges reach almost across the width of the door. The middle hinge can also span a plank door or a panel door with a lock rail, but he cautions, "Hinges should not be attached to panels, which must be free to move with changes in the weather."

In the last two years, Bob has made 17 doors and their hardware. This experience has allowed him to experiment with a variety of ways to articulate the hinge on the pin. However, he usually offsets the hinges by overlapping one hinge plate above the other on the jamb or makes his own three-knuckled hinges (with a table saw or band saw).

Whichever method he chooses, Bob begins making each hinge plate from what is literally "an uncarved block"—seasoned wood up to 6 inches thick or laminated wood stock. He planes the surfaces so that they are absolutely flat. Then, when he is sure his placement is perfect, he drills the hole for the hinge pin. He tests the fit of the hinge pin (usually a maple dowel bought locally) by aligning both hinge plates and dropping the pin through. If the hinge pin is to project above the hinge plates, he caps it or fits it with a cross-pin. Then, with gouges, rasps and files, he carves his form.

When Bob has finished the hinges, he mounts the appropriate plates on the door with ¼-inch lag screws at regular intervals. He counterbores holes for each screw by drilling ¾-inch holes in the hinge plate, and then continuing to drill at the center of each hole with a 9/32-inch bit in the drill chuck until the drill pokes through. When all the holes are drilled, he slips bottom, middle and top hinges over a 6-foot wooden

dowel or mild steel rod to give him perfect vertical alignment. He lays the dowel and hinge plates at the edge of the door and puts the hinge plates in proper positions along the rails. Then he puts lag screws in the holes and drives them into the door with a socket wrench.

"In some designs," Bob says, "I leave the holes unplugged, allowing the attachment technique to show. Alternatively, you can fill the holes with a standard washer and commercially available wooden plugs; or, with a ¾-inch plug cutter, you can make plugs of a contrasting species or cut plugs from the same stock used to build the hinges." Whether he leaves the plugs protruding or finishes them flush with the hinges, Bob avoids gluing them in so that the hinges can be easily removed later should the door need refinishing.

To mount the door on site, he shims it into proper position. He joins the hinge plates attached to the door with those to be mounted on the jamb by dropping in the pin. Then he drills holes through the remaining hinge plates, through the doorjamb and into the rough framework of the house. In these holes, he sinks more ¼-inch lag screws—long enough to anchor the hinges securely. Then he may or may not plug these holes (according to the design) before finishing them. Finally, with the door slightly ajar, he puts a flat-head wood screw through the knuckle of one hinge plate and into the hinge pin of each set of hinge plates. This prevents the pin from migrating up or down when the door swings open.

He cautions those who would make their own wooden hinges that doors fitted with such

hinges do not swing open as quickly as those fitted with steel hinges. In addition, they sometimes develop "healthy squeaks," which can be annoying. The noise can be quieted with a dusting of talcum powder or a squirt of a light machine oil (like 3-in-1 Household Oil).

Bob plans to improve his hinges (and solve the squeak problem) by lining the holes with a metal sleeve. This will not change the looks of his handmade hinges at all, but it will make his doors swing more freely.

Even more than hinges, Bob likes to make door handles. He sculpts birch into a shape that looks like a sliver of bone, applewood into an omega, mahogany into a tricornered lever that nicely fits the hand. He believes, "Handles must meet more than a test of the eye. They must be comfortable in the hand and functional."

Once he has decided on his design, he works again with "an uncarved block," roughly shaping it with a band saw, and then whittling away at it with gouges and rasps until it has assumed its intended form.

He finishes the handle with rasps, then files, and finally sandpaper. With the coarser grades of sandpaper, he uses sanding blocks to keep the surface uniform. Once he has progressed to 100- or 120-grit sandpaper, he trusts his hands. When he has reached 220-grit, he wets down the wood with water to raise the grain, lets it dry, and resands. If he feels even finer sanding with 400-grit sandpaper and above is required, he may wet-sand with a tung oil or penetrating resin finish, applying the oil, sanding the surface while wet, and wiping off the excess.

Because of the way Bob made this screen door hinge, above, left, *he can lift the whole door out of the door frame when winter weather requires a storm door. Both the screen door hinge and the shop door handle,* above, right, *are carved from maple.*

Bob often spends as much time finishing his doors and hardware as he does making them—as the results show. However, he is not immune to making finishing mistakes. The "bone sliver" birch handle had to be finished twice because the Danish oil that he used to wet-sand it did not penetrate evenly, giving the handle a splotchy appearance. He was very discouraged—but decided to file and sand it down to beneath the depth the Danish oil had penetrated and try again, this time preceding the finish with a sealer. To his relief, his second experiment worked.

After years of making handles, he has discovered that there are some woods that are "naturals." Apple and yew are polished to a nice patina by constant handling; maple and birch tend to get grubby, although they make beautiful handles if given the protection of varnish.

Bob attaches the handles to the door with dowels and glue, or with counterbored lag screws. So far, all of his handles have been fixed, and sometimes mounted on a matching push plate. Recently, however, Bob met up with a machinist who agreed to adapt the inner workings of a standard entry lockset to fit Bob's handles, so in the future Bob will be making movable door handles that can be securely locked.

A few years ago, Bob Jepperson made a tough choice. He decided to risk job and financial security to live close to the land. He didn't expect to make a living doing what he loved. In fact, he didn't know how he would make a living. He says softly, but firmly, "I didn't give up my job and my life in the city to do woodworking. I gave them up for a more self-sufficient life, and woodworking just came naturally as a part of that."

Opposite: *The applewood omega handle has been polished by use to a satiny patina over time; the sliver-shaped birch handle requires more maintenance.*
Left, top: *Bob carved an intricate maple molding to emphasize the beautiful simplicity of this panel door.*
Left, bottom: *Another birch handle acts as a push bar on this attractive passageway door.*

Aj Darby

AN ART NOUVEAU PANEL DOOR

Aj Darby came to the United States from Britain in 1974 with lots of talent, training and experience, but little money and no tools. He'd studied architecture and graphic arts at the University of Birmingham, trained as a computer programmer, owned a disco, managed a graphics arts business. He'd salvaged doors off Victorian houses and restored them, made jewelry, designed stage sets. He intended to continue making jewelry, but without the tools for metalsmithing, he had to give up this ambition. Instead, he took small construction jobs and did a stint in a woodworking shop until finally he was able to go on his own, alternating between carpentry and commissioned sculptural cabinetry. Now he works as a subcontractor, designing and installing decking and laying out garden walkways or building the occasional, extraordinary door on commission.

"I don't really look at my work from a cabinetmaker's point of view," Aj explains. "I'm not into perfect joints—partly because I don't have the tools to make them. I get this image in my head, and I put it out." Aj looks at each door he makes as a piece of sculpture. If the design is pleasing to see and to feel, and if the door swings smoothly on its hinges, and is finished to last, then he is satisfied. He doesn't waste time worrying about the looks of parts that are hidden, since they do not detract from the beauty or strength of the finished door.

The panel door shown at left was commissioned by a couple who had seen some of Aj's other work. They had remodeled and enlarged their house and wanted to dramatize their new entry with a door that would be "a little out of the ordinary." Since they had a scarlet macaw, Aj drew up a design that incorporated stained glass sidelights showing the macaw sitting on a branch. He carried the design into the door by making the branch curve and extend around the panels as decorative molding in an art nouveau style. The couple okayed the scale drawing, and a local craftsman went to work on the stained glass windows while Aj made the door.

First he enlarged his scale drawing, using a grid system, to give himself cutting patterns for his lumber. Then he bought varying widths of 1-inch-thick kiln-dried redwood, surfaced on all four sides, from a local lumber mill. "This is an expensive way to buy redwood," he says, "but it does provide me with a high-quality material that requires the minimum of tools to convert it to the finished piece."

To rip the lumber, Aj borrowed a friend's

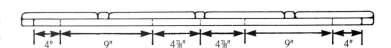

Cross section of first two layers of Aj Darby's door assembly, using 1'' redwood (actually ⅝'' thick) for each layer.

Interior view of panel door made by Aj Darby.

table saw. The rest of the tools he needed he owned—jigsaw, hammer, Surform (a kind of plane with a rasplike cutting surface instead of one smooth cutting blade), utility knife, belt sander, sandpaper and worktable.

He made the door in three layers, assembling it in such a way that the core and exterior layers appear as panel and frame. He cut the redwood to fit the height of the door opening, and to varying widths. He assembled them on his worktable in this order: 4, 9, 4⅞, 4⅞, 9, and 4 inches. He used marine glue along all the joints, clamped them tight with bar clamps, and weighted the core layer down with rocks to prevent warping as the glue dried. When it was dry, he cut two more boards to span the top and bottom of the door, and glued these in place. By doing this, he minimized the amount of exposed end grain, making it much less likely that the core layer would ever warp from moisture absorption once the door was in place.

Next he cut boards for the interior of the door, wide enough to overlap the joints of the core layer. To give the appearance of molding, Aj cut very narrow strips of redwood that he had rescued from a chicken shed. "I'm not sure what chemical changes the floorboards went through over the years beneath the chickens," Aj admits, "but they had a *wonderful* purple-red color." He rounded the edges of these boards, first with the Surform, then using sandpaper. He began sanding with 60-grit, followed with 80, and finished with 100 or 120, depending on the hardness of the section he was sanding.

When at last all the edges were smooth, Aj glued the interior and the core layers together.

He beaded glue on the surface of the core layer and spread it around with a scrap piece of wood; he also glued along the joints of the interior layer, alternating purchased with salvaged redwood for color. The marine glue he used stayed tacky long enough that he could shift the boards around to make all the corners square. Again, he clamped across the width of the door to make the joints tight and weighted the door so it would dry level overnight.

Next he cut the opening for the window that was to go in the center of the door. He made a starter hole with a drill and then cut the opening with a fine scroll blade in the chuck of his saber saw. He cut the hole large enough that the glass could be installed from the interior, with the window trim on the exterior serving as a stop.

To make the exterior layer of the door, Aj traced the individual design elements of the full-scale drawing one by one, to give himself a cutting pattern for each piece. Then he placed his tracing paper pattern on the wood, and he checked the direction of the grain to determine the best way to position it. (In general, it is best to lay the pattern out so that the largest dimension of the pattern follows the grain.) He transferred the pattern to the wood by tacking the tracing paper in position, drawing side down, and running a pencil point over the outline. Finally, he cut out all the shapes with his saber saw.

He checked the fit of all the pieces by assembling them on top of his full-scale drawing on a flat surface, and adjusted as needed. Great precision was not necessary. He explains:

These close-ups show details of the redwood molding Aj Darby applied to the front of his entry door.

"Because all the pieces had rounded upper edges, they didn't have to fit together exactly. Anything up to a ⅛-inch gap between the pieces was entirely acceptable and virtually invisible in the finished product."

Aj shaped the decorative molding pieces roughly, using his Surform on the straight edges and a utility knife on the curves to make them look as though they overlapped and intertwined. He left the four outer frame pieces — stiles and top and bottom rails — square ended for butt-joining.

He hand-sanded all the pieces with several grades of sandpaper, beginning with 60-grit. He says, "I used a toweling action across the grain. It was the fastest way to remove the marks left by the Surform and the utility knife." Although Aj was able to clamp the larger pieces steady for sanding, he had to hold the smaller pieces in one hand while he sanded with the other — all the way from 60-grit up to 120.

Next Aj sanded the "panel" areas of the core layer with 60-, 80-, and then 100-grit paper. After wiping away all the dust with a tac cloth, he glued the exterior layer to the core layer, again using clamps and weights. After a couple of days' drying time, he sanded away the glue lines that showed on the outer edge of the door.

To install the window, he turned the door over so the interior side was face up. He made the inside window trim and sanded it. He installed the glass with glazing points and putty, then glued on the inside trim.

Finally, when he was certain all the glue was dry, he applied several coats of a clear penetrating resin finish to both sides of the door to protect the wood without hiding its texture. The product he used, Varathane Oil and Sealer, is no longer on the market, but there are many finishes like it that are easy to apply simply by following the manufacturer's directions.

The finished panel door was delivered to its new home. A carpenter hung it in its frame and added hardware. He also installed the stained glass sidelights. Stangely enough, the couple who owned the house decided they preferred clear glass sidelights, so the stained glass panels were removed. Now the door Aj made gets the full attention of everyone who comes up the walkway.

Most people find out about Aj by word of mouth, but his reputation has not yet spread far enough that he can afford to set up a woodshop with a full range of tools, hire an apprentice and make doors full time.

Aj Darby's circumstances are hardly ideal to do fine woodworking, but because of his resourcefulness, he has been able to make some extraordinary doors — doors that give him satisfaction and the incentive to continue to use his time to express his artistic talents.

Donald Pecora

THE WARM DOOR

Donald Pecora graduated from a well-known university in 1974 with a B.A. in international relations, but gave up his chance at a foreign service career to become a woodworker in Alstead, New Hampshire. From the way he tells the story, it wasn't even a difficult decision. Memories of the summers he'd spent in Alstead as a child drew him back with an irresistible power.

"I think I was seven years old when I went down to Heman Chase's Saturday woodworking class for the first time. I made a wooden cart, and I had one person put the wheels on it and another person put the boards down. Heman told me that I should come back the next year when I was a bit older and didn't need quite so much help."

Don did go back the next year, and for many years after that. As he grew older, he helped Heman teach woodworking to the younger children. And when

70

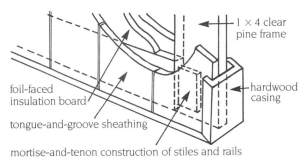

Cross section of a Warm Door showing core construction.

1 × 4 clear pine frame

hardwood casing

foil-faced insulation board

tongue-and-groove sheathing

mortise-and-tenon construction of stiles and rails

he returned to Alstead as a grown man, he knew how he wanted to spend his days.

When he discovered that someone else had started a woodworking shop in the area, he told them that he would bring his wood lathe and work for nothing just for the privilege of being in the shop. He held down part-time jobs as a waiter and as an orderly in a nursing home to keep food on the table while he learned his craft. He kept his eyes and mind open for a wood-working item that he could make his specialty.

He made his first insulated door in 1977, as an anniversary present for his parents. After he'd watched the door weather a few severe New Hampshire winters, he accepted a commission to build doors like it for a timber-frame house. Then, with the help of his wife, Anne, and a friend who was a graphic artist, he put together a brochure featuring photographic examples of the work he'd done on the timber-frame house and an illustration of the construction of the Warm Door (a registered trademark). Orders began to pour in. Donald Pecora was suddenly in business for himself, with a woodshop of his own, a couple of apprentices, an accountant, and no regrets.

He says about his company: "We don't have one person who does all the milling and one person who glazes windows all day. I try to see that each door is taken from step one through completion by one person. So there's a lot of that person that goes into the door." Don encourages each woodworker in his company to be a crafts-man rather than an assembly-line worker — even though the Warm Door is produced in standard as well as custom designs.

Don uses lumber taken from New Hamp-

shire forests. After years of sorting through lumber at commercial lumberyards and coming away with little that satisfied him, he made arrangements with a few of the area sawmill foremen to go out into the woods with them and pick the pine trees he wanted logged for his doors. He always tries to select trees that have grown straight and are free of stresses. Then he rough-mills and kiln-dries the lumber at his workshop. He pays for the privilege of selecting his lumber, but thinks it's worthwhile: "It's just a pleasure to go up to the wood rack and pull out board after board and say, 'Oh . . . this is a nice one!' "

Each Warm Door is made very much like the one that Don presented to his parents: The woodworker making the door chooses pine boards from the wood rack for the internal frame and sheathing, and oak or cherry for the hardwood casing that will edge the finished door. He cuts mortises and haunched tenons in the 1 × 4 stiles and rails of the internal frame and assembles it, wedging the tenons and gluing the joints.

Then the woodworker fits foil-faced insu-lation board within this frame. He applies a sheet of clear plastic over the insulation board on the door's interior to act as a vapor barrier. These two measures help boost the insulating value of the door to R-10, 3½ times that of a solid wooden door.

The woodworker applies 7/16 by 6-inch tongue-and-groove sheathing to both sides of the door in a pattern of the customer's choice. Finally, he edges the door with 2¾-inch-wide strips of oak or cherry, grooved to receive the 1⅞-inch thickness of the door, and mitered to make neat corner joints. (The strip of casing

The Warm Door comes in several standard styles: The European T, above, and the European and the Country Dutch, opposite.

foil-faced insulation board, and frames it with more of the hardwood casing. Then he lines the window casing with closed-cell polyvinyl weather stripping. He makes a separate frame for the sash, and lines it with the same polyvinyl weather stripping, then fits the layers of glazing into the frame and secures them with beveled molding strips, tacked in place as stops. He uses no putty and no silicone. This allows any condensation that develops between the layers of glazing to dissipate.

Don is especially proud of his wife, Anne, who gave up her potter's wheel to learn about stained glass. The Pecoras spend many Saturdays together at the shop, where Anne cuts glass for leaded windows. Very often, she works personally with a Warm Door customer to help in the selection of colors and designs.

Don offers all Warm Door windows with glass or acrylic glazing. "Some people think glass is prettier," he says, "and I agree — but the acrylic is a lot warmer than glass. It just doesn't transmit the heat or cold the same way. It reduces sound quite a bit, too. And it can be maintained scratch free if you are careful to wash it only with gentle soap and water."

The Warm Door comes prehung in its own jamb for installation in new construction or by itself so that it can be hung in existing structures prepared for the 2¾-inch-thick door. Whoever makes the door makes the jamb, too — of sturdy 1¼-inch pine — and fits it with a hardwood sill. To seal any cracks around the door that might allow infiltration, the wood-

that fits along the lock side is also beveled 2½-3 degrees to make the door open and shut smoothly and without resistance.)

The woodworker beads glue inside the casing along the hinge side of the door to add structural strength where it will be most needed, then dabs it on the end grain to reinforce the miter joints. He drives long wood screws through the hardwood casing into the core and covers them with wooden plugs.

If the door is to have a window, the woodworker next cuts the opening through pine and

worker uses a combination of brown vinyl-covered neoprene foam weather stripping and a doorstop. Finally, he tacks short strips of a fuzzy wool-pile-like weather stripping in the corners of the door frame.

Don believes the effectiveness of an insulated door is maximized by the weather stripping that goes around it, so he has tested many brands to learn which are most durable and have the best memory (the ability to return to original shape and size after compression). He is especially proud of the closed-cell polyvinyl strips he uses to weather-strip windows. "This stuff is really quite amazing," he says. "If you compress it 10 percent, it's completely airtight and watertight." Meanwhile, he keeps his eyes open for new weather-stripping systems that could improve his own product—and he tests the airtightness of every prehung door before delivery by blowing compressed air (at 120 pounds of pressure per square inch) along the perimeter seal of the door in the door frame. A co-worker stands on the other side of the door to check for air leaks.

How the door is finished is up to the customer. If he requests rough-sawn pine, then the woodworker will go over the rough-sawn sheathing of the door with a Surform and a heavy wire brush to give it the desired texture. If the order is for a smooth finish, he will plane the door down and run it through a modified platform sander, which sands it inch by inch to a uniform cut. The woodworker goes over any areas of raised grain that remain with a cabinet scraper (a hand-held tool of high-grade metal with a raised burr for producing very fine shavings; it is especially useful on hardwoods).

All doors, doorjambs and sills are treated to a bath in Cuprinol, a chemical wood preservative, before being finished.

Don tests the durability of the various finishes he uses by coating scrap pieces of pine with the products and putting them outdoors, where they are subjected to the sun's ultraviolet rays and the region's heavy snows, gusting winds and temperature extremes. The ongoing tests allow him not only to gauge each product's performance, but also to recommend a maintenance schedule to his customers.

Don's company now offers windows as well as several styles of doors, distributes the weather-stripping products they use (which are not generally available at hardware stores) and markets a Warm Door kit that supplies lumber, insulation and weather-stripping for those who would rather make their own door. In addition, he says, "We're willing to do any carpentry or cabinetmaking that comes into the shop. I like to think of us as filling a niche here in the community for a general woodworking shop. If we can help anyone out with anything in any way, we don't say no to him." It seems that Heman Chase taught Donald Pecora many lessons.

Adi Hienzsch

CARVED DOORS

The area around North Bend, Washington, bears small resemblance to Garmisch, Germany, where Adi Hienzsch was born and grew up, or to Oberammergau, where his friends, professional wood-carvers, taught him basic woodcarving technique. The mountains are high, but do not fill the sky like the Bavarian Alps; the snows of winter are deep, but not so deep that they never melt. Yet Adi, in his patient, unassuming way, has changed North Bend and the nearby towns by the Old World touch he gives to his work. Every one of the doors he has made for private homes and businesses in the area expresses the values of the German wood-carving tradition.

Adi learned to carve 25 years ago. As a member of the mountain police, he often had to stay at posts high in the Alps for a week or two at a time. Although he was occupied with patrolling the Bavarian-Austrian border during the day, his evenings were free for other pursuits. He enjoyed painting as a hobby, but found it inconvenient to lug oils, mixing palette and canvas in his backpack. His friends convinced him that a couple of wooden blocks and a carving knife would be easier to carry, and told him that the souvenir items he could make with a few quick strokes of the knife could provide a nice second income. After a few informal lessons, he began carving small items, and soon had a steady stream of orders to fill.

When he came to the United States in late 1963, it was his experience on the mountain police that helped him find work. He worked during the snow season as a professional ski instructor, continued to carve in his leisure time, and saved his earnings toward a down payment on a woodworking shop. When he finally did go into business, he decided not to do the small souvenir items that industrial woodcarving machines can mass-produce so quickly, but to do only large-scale items like murals and doors, which no machine can duplicate. For the last 19 years, he's been quietly making carved entry doors in his own workshop and showroom, Edelweiss Woodcarving Studio.

He begins his doors with a trip to the local lumber mill to select a lot of Alaskan yellow cedar. This wood, which is native to the Pacific Northwest and British Columbia, is particularly noted for its resistance to decay, so it is especially good for exterior use. In addition, this close-grained wood is easy to carve and grows harder with age, making all the handwork more likely to endure. The Japanese, who have little hardwood in their own country, have been buying up the best yellow cedar, sometimes using it right away but more often simply storing it in cold water or brine for future use. As a result, Adi has to buy about three times as much rough-milled wood as he actually uses for each door so that he can select pieces that are clear of knots, flat, and without end checks. All of the cedar is kiln-dried at the lumberyard, then allowed to air-dry under a covered shed for at least three or four months before it is used.

When Adi first began making his carved entry doors, he assembled the door slabs himself. Now he believes he can make better use of his time and talents by having a local millwork shop assemble the doors, while he concentrates on carving. He does, however, oversee the layout and assembly operation.

Adi has the millers surface and join the boards to make them uniform in thickness (usually 1¾ inch thick, standard size). Then he has them lay out the wood so that vertically grained heartwood, cut from the middle of the log, joins more vertically grained heartwood, with the curve of the ring pattern reversing in direction at each of the joints. This layout provides good structural stability in the finished door, and the consistent grain direction is advantageous for later wood carvings.

For a standard door, 36 by 80 inches, the millers butt-join seven boards, each no more than 6 inches wide. Adi says, "Often people ask, 'Where do you get a piece of wood this wide (36 inches) for a door?' But I have to say, 'Even if I could get a monster like that, it would not

Adi Hienzsch designed and carved this door for an American Indian who wanted those who entered his house to understand his love of nature.

76

be good to use. The ability for it to stay in place is much less than with smaller boards.' "

Once the boards are numbered for assembly, the millers glue them together with a waterproof glue (such as Weldwood's resorcinol), clamp them with bar clamps, and let them cure. When the door is dry, the millers sand it on an industrial belt sander to an even texture. Finally, they install a T-iron (a standard item available from any steel wholesaler or welding shop) across the top and bottom of the door to do the same job a steel dowel would do in a manufactured door.

Adi says that a solid-core flush door, to which a 1-inch-thick panel is applied, can provide an equally satisfying carving surface. He suggests that those who would like to try making and carving their own door buy a solid-core flush door, and build up a panel of the same kind of wood as the veneer laminated to the surface of the flush door—birch or oak or mahogany all look nice and carve well. Lay out the boards with care, and butt-join them with glue, under pressure of bar clamps. After the glue has dried, trim the panel to exact size, and plane and sand it by hand.

With the solid-core flush door resting on two sawhorses, apply glue to its surface. With the carving panel on your worktable, spread glue on its underside. Then mount the carving panel on the door. (The carving panel need not be cut to the full dimensions of the door but can be inset to accommodate hardware. If you inset the panel, mark its position on the door before mounting it and take care not to spread glue outside the marks.) Clamp the panel tight against the door, using C-clamps to apply pressure and hardwood boards spanning the panel to distribute the pressure evenly across its surface. Protect the door surface from marring by placing wooden blocks between the C-clamps and the underside of the door.

After you have carved your panel, you can further secure it to the flush door by nailing paneling nails into the wood at 6-inch intervals along the panel perimeter. Adi says, "In a carving, nails are easy to hide. Set them deeper (with a nailset), and cover with wood filler. You will never see them after finishing."

Adi plans the design of the carvings on his doors in discussions with his clients. The door shown at left was carved for an American Indian. The client wanted Adi's realistic style of carving rather than an Indian totem, but did want the door to express the reverence for nature that is so much a part of his heritage. He chose the subjects for the carving—the eagle and the fish. Adi was pleased with the end result. He says, "You see the door, and are right away touched by the feelings of the house owners."

Adi suggests that anyone wanting to try his hand at carving a wooden entry door not duplicate any existing design, but create his own. It is the woodcarver's special joy to see the work of his own imagination take shape under his hands. "But," he cautions, "carving requires expertise; plan the design to your level of skill."

When Adi has an idea in mind for the carving of a door, he makes several sketches on tracing paper, reduced to a scale of 1:10. Sometimes he lays a sketch showing foreground over another showing background, to see if all the design elements are in proportion. Differences in proportion and amount of detail set the foreground off from the background, giving the effect of depth.

When Adi finally comes up with a design he feels is pleasing, he projects and enlarges the layered sketches directly onto the surface of the door, and retraces the image with a pencil as a guide for his carving.

For those who don't have a projector, Adi suggests simply enlarging the 1:10 grid to full scale and then transferring the design to the solid wooden door surface with carbon paper.

Finally, it is time to carve. This is accomplished by first making stop cuts (cuts straight down into the wood to sever the fibers of the grain and establish the depth of foreground, middleground and background) following the outline of the design, and then slicing cuts (cuts angled toward the stop cuts to lift away wood chips). Although Adi often makes his stop cuts with a router, he advises against this practice for all but experts. Chisels and gouges are the woodcarver's stand-bys.

Adi begins carving with the door supported on two padded sawhorses, set at a comfortable height. He uses his router, fitted with a V-grooved bit, to outline the rectangular perimeter of the design, using a straightedge clamped to the door as a guide. With a straight bit, he makes freehand stop cuts around irregular shapes in the design.

If you would like to try using a router to make stop cuts, practice on scrap wood, learning how the router cuts with, across and against the grain. When you begin to carve the surface of your door, hold the router back from the pattern outline so you can later smooth the edge with hand chisels and gouges.

Whether you use a router or chisels and gouges to make your stop cuts, watch the depth. It is better to work the cut deeper little by little than to make the first stop cut so deep that the design suffers. More wood chips can always be carved out, but once removed, wood cannot be put back.

Once the outlines of the elements in the design are marked by stop cuts, Adi removes the background stock. He uses wide gouges (curve-bladed chisels) to carve out waste wood to a shallow depth (⅜ or ½ inch), and then narrower gouges and chisels to take the background stock to its finished depth and create an appropriate texture.

Adi recommends that particular care be taken when carving with a chisel or gouge in the direction of the grain. Under all but the gentlest, surest touch, the wood will split beyond the intended cut. Carving across the grain is easier because the response of the wood is more predictable.

The process of carving a design proceeds gradually from outlining and rough shaping to precise definition of shapes and textures. Although some surface detail is carved while the door is still on the padded sawhorses, Adi finishes his carving with the door in an upright position. He does this to get a feeling for how the door will look on a house. He says, "You will quickly see the wrongs of perspective which you could not notice while working with the door lying down.

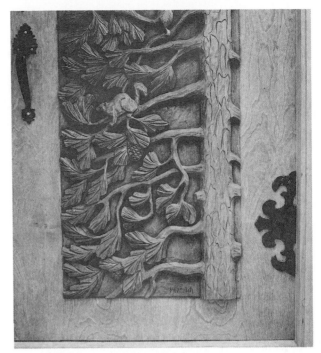

Use your eyes and sharp chisels to make final corrections and overall touch-ups. This will lead you to the satisfactory result you expected from your imagination."

Adi finishes the door by lightly sanding the background carving to remove rough spots and raised fibers, which would show up as dark spots when the door is stained. As for the chisel marks, he says, "The chisel marks are my handwriting. They show the carving was truly done by hand. Why destroy them?"

After sanding the uncarved areas of the sides, top and bottom of the door to a smooth finish, he vacuums the door and wipes it down with a tac cloth to remove all dust. He supports the door on padded sawhorses while he stains and varnishes it with Flecto products. He says, "I get the right penetration control with this stain. I can stain the whole carving and still control light and shadow by wiping off excess stain using a clean rag."

After staining it, he varnishes the door. He uses a small paintbrush to reach into all the corners and irregular surfaces of his carving, being careful not to apply the finish too thickly. He uses a foam brush to coat the flat surfaces without leaving brush-stroke marks. He also seals the bottom and top edges of the door, protecting the iron T-bars from rust.

And then? The door, with chisel marks that are the handwriting of a master woodcarver, leaves his studio for the home of a storekeeper or a teacher or a factory worker. And the people who see it know a little bit more about Old World craftsmanship.

Adi Hienzsch's carving is precise and beautiful whether he uses it to create a repeated motif or to depict a whole scene.

*James Parker and
Bob Richardson*

BARNWOOD DOORS

The Fenn Gallery of Western Art in Santa Fe, New Mexico, is the largest gallery of its kind, and only one of several buildings on a private estate whose owners are devoted to the support of western art. Both James Parker, a full partner in the custom-building firm, Cerros Construction, and Bob Richardson, a craftsman woodworker, have made barnwood entry and passageway doors for the buildings on this estate. Even the casual visitor is impressed with the uniqueness and beauty of each of these doors. Yet James was asked to turn out 20 simple, rustic doors as quickly and cheaply as possible, and Bob was told to take his time and lavish the craftsman's attention to materials, construction and design on each door he made. What, then, are the subtle differences between materials and methods James and Bob used, and the effects they achieved?

James Parker was working as a carpenter for another contractor when he was asked to build doors for buildings at the Fenn Gallery estate. "The owner, the architect, the contractor and I each had a part in deciding on not only the placement and sizes of the doors, but also the style and cost," he says. "We decided that it would be difficult and expensive to find weathered wood in sufficient quantity to be able to select for straightness and soundness and consistent thickness, so we decided instead to reproduce the old look." To accomplish this task, James rethought the basic construction process for panel doors. Instead of cutting the stiles to run from sill to head jamb and to receive the tenons of the rails, he cut the rails to span the width of the door, and grooved them full length, ¾ inch deep, to receive a plywood panel and the tenons of the stiles. He also grooved the inside edge of all the stiles the same depth to receive the plywood panel. Using construction grade pine 2 × 4s or 2 × 6s for the stiles, top and lock rails, and 2 × 8s or 2 × 10s for the bottom rail, he was able to fabricate the frame

These barnwood doors by James Parker, a custom builder, opposite, *and Bob Richardson, a fine woodworker and cabinetmaker,* left, *were made for the same estate, but still reflect the individual differences in the doormakers' work methods and design styles.*

These details from the central panels of James Parker's doors show how imaginatively he varied the barnwood applique to create one-of-a-kind designs.

members for all 20 doors at one time. He also cut lumber for the jambs. He cut panels to fit the doors from ⅝-inch C-D plywood.

He assembled the parts to check the fit, then glued and clamped the doors together. "We used a yellow glue for interior use— Franklin's Titebond," he says. "So we had to assemble and clamp the stiles, rails and panels in the space of ten minutes. We had two people working on the door at this stage, so that the parts could be tapped together with a rubber mallet and aligned before the glue set."

Perhaps James's most original contribution to the project was inventing a technique to reproduce the look of weathered wood on the frame members of the door blank. Essential to the success of his experiment was the use of scrap pine pieces to test the effects of applied pressure, burning and stain colors. First, he roughed up the frame members using a hand-held electric grinder with a coarse wire wheel in place of the grinding disc. Since stiles and rails were of soft pine, this treatment cut fairly deeply, raising the grain pattern in places. He did not rough up the plywood panels.

Next he scorched the stiles and the rails black with an oxyacetylene torch to simulate the oxidation of weathering. He advises against doing the scorching in the woodshop, however, because even a small amount of sawdust can fuel a fire.

When he had finished scorching stiles and rails of all the doors, he used the wire wheel to remove the charred wood in the softer areas. He allowed the denser-grained, pitchier areas to remain darkened. The effect of unevenly raised

82

grain is very like that of natural aging.

To antique stiles and rails, James used an oil-based stain from Minwax called Driftwood, and varied its flat gray tone by adding small amounts of ochre, cream, dark brown and black oil-based paints. He mixed the stain and paint just as an artist would mix color on his palette — with some experimentation and some intuition. He tested his colors on scrap wood before brushing them on the frame members. "Remember that what we were trying to do," he says, "was match the character of the old wood rather than just one piece in particular." To achieve the effect he wanted, he even tried applying one coat of stain, and then rubbing dabs of black and white paint on different areas of the stiles and rails to imitate the uneven bleaching of the desert sun.

Since most of the door blanks were similar, the truly creative design work didn't occur until the doors were ready to be appliqued. James made up sketches detailing the applique patterns on the panels for each door, and then began cutting barnwood to fit. He used a table saw to resaw the wood to 3/8-inch thickness and to cut it to the necessary length and width. He found that the old wood was quite difficult to work with. When he attempted to resaw it down to a veneer thickness, it often fell apart. Sometimes it was so dry and brittle that it had no structural strength. He rejected the worst of the old wood for this reason.

He applied his barnwood to the plywood panels, usually working from the bottom up and from the center outward. To get curved pieces to fit within the stiles of an arched door, he laid them on the panel surface against the curve of the door, scribed his cutting mark and used a jigsaw to cut them to fit. If a barnwood pattern piece had a knothole in it, he painted the plywood area that would show through behind it black.

He glued all the pattern pieces to the panels with interior glue, and clamped the doors with bar clamps. He also used small brads to secure the barnwood pattern pieces to the plywood panels. He tried setting the brads and found that this marred the wood, so the flat, stainless steel heads are visible. He would have preferred to make a face clamp (a piece of plywood that could be put down on the door and clamped in place) for each door and to do without the brads, but that would have taken too much time.

It was not necessary for James to apply any additional finish to the doors he'd made. By selecting barnwood that was level and free of dry rot, he lessened the chances of warpage along the panel surfaces. His antiquing process sealed other parts of the doors from changes in moisture content that could cause warping. And the dry, hot New Mexican climate itself acts as a friend, helping to preserve the structural form and surface beauty of the doors. For all these reasons, the doors James Parker designed, built and installed at the Fenn Gallery estate are likely to look and perform well for a long time.

The owner of the Fenn Gallery is confident of the quality of James's work — so much so that he recently sought out James for a special commission. "After I started my own company," James says, "the owner of the gallery came to me and asked me to do a guest house, library and guest apartments. I hired Bob Richardson, a full-time cabinetmaker, as a subcontractor to do the doors for these buildings. He adapted some of my procedures to his own style. It's been an ongoing project. . . ."

James explained the antiquing process he had devised and invited Bob to use the same technique, but Bob wasn't comfortable with the idea of working with a hot torch. He had collected enough old wood over a period of years to build all the doors that had been commissioned, so he decided to use these materials rather than go through the "antiquing" process. In addition, he planned on varying the applique process with subtle touches of his own.

Wood gathered from barns, sheds and corrals tends to be rather dirty. Bob describes it as " . . . pretty ratty looking — covered in pigeon droppings, dirt and sand — and there are always a few nails that have gone undetected, just set to destroy all those fine, expensive carbide blades." So Bob's first task on this project was to go to town and buy some inexpensive (and expendable) carbide rip blades for his table saw. Then he looked through his lot of barnwood for 2-inch stock that was free of twists, bows and crooks.

Since the salvaged wood was not up to the doormaker's exacting standards, he resawed it to a consistent thickness. When necessary, he resawed it in half, sandwiched another piece of wood in between and glued the three together to get a piece of lumber of the dimensions he needed.

Working with the weathered wood was not easy. "The wood I had stocked up, like most of the wood used for building in New Mexico, was pine. The knots had dried years ago. Sawing through them was like cutting through glass. Bits and pieces flew everywhere. I *had* to wear my safety glasses!"

Like James Parker, Bob made his panel door frames using mortise-and-tenon joints. He grooved stiles and rails to a ½-inch depth to receive the ½-inch plywood panels. He assembled the pieces, applied glue, and clamped the doors together square and true. Unlike James, he didn't make sketches for the applique patterns on paper. With the doors propped against a wall, he drew his design directly on the plywood panels.

Cutting barnwood for the applique sent him back to his table saw. He rough-ripped 1 × 6s, 1 × 8s, and 1 × 12s to 1¾-inch width, then resawed them to a uniform ¾-inch thickness. This showed all the remaining nails as bright silver, so he pulled them out. Then he passed the lumber through a planer to reduce it to a ⅝-inch thickness, ran it across the joiner to get straight edges, and then went back to the table saw to rip it to a consistent width. This made the wood straight and evenly dimensioned.

He cut his applique pieces to size, rabbeting one end to set off each joint. He highlighted the rabbet by painting it turquoise or a mix of turquoise and black. Then he glued the pieces in place with a construction adhesive, following the design drawn on the panel, and clamped the pieces down while the glue dried. He used brads only on those applique pieces that didn't adhere perfectly.

Craftsman Bob Richardson notched and painted the joints between his applique pieces to give his barnwood door, left, a formal elegance, but he also made doors for the estate that are more "fun." Just to be different, he paneled the "saloon-style" doors, above, with river reeds.

The resulting doors are very like the ones built by James Parker. James was told to keep costs down and thus worked with as many inexpensive materials as he could find; Bob had barnwood on hand and was able to discard 50 percent of it as unusable, selecting for lack of defects and warp, clarity of grain, and superior color. James's doors show the powers of imagination and invention; Bob's the value of patient craftsmanship and personal style. One is not better than the other—both men got beautiful results working within the design parameters they'd been given.

FLUSH DOORS WITH FLAIR

When Lesta Bertoia was 16, she worked for her father, the sculptor Harry Bertoia, building packing crates to ship his artworks all around the world. As he worked with a welder's torch on his metal sculptures, she worked beside him with hammer and saw. Learning on her own to measure and cut wood accurately and to work with power and hand tools made her confident that she could succeed at woodworking. Since then, she has made furniture, lamps and children's toys. Her boomerangs have been her most successful venture, selling well at craft fairs all over the country.

In the past year, however, Lesta has given up the production of boomerangs to work with designers and builders on an underground, passive solar house and separate studio for herself and her family.

Lesta prepared for her involvement in the project by enrolling in a three-week inten-

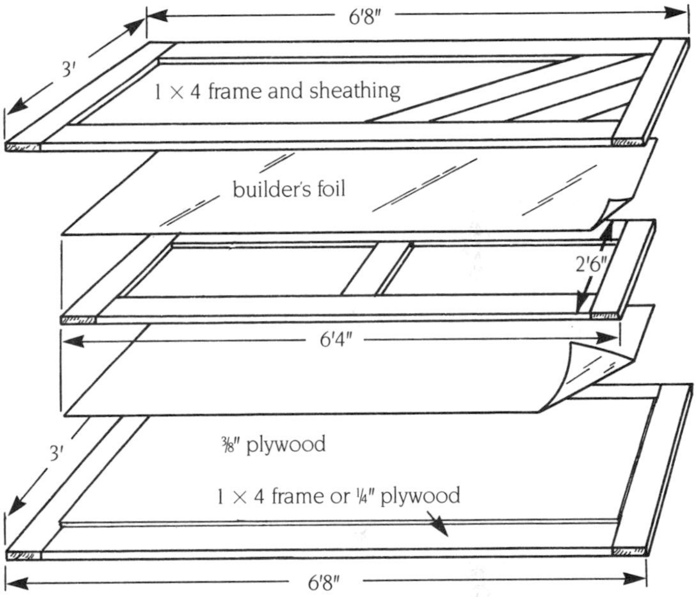

6'8"
3'
1 × 4 frame and sheathing
builder's foil
2'6"
6'4"
3'
⅜" plywood
1 × 4 frame or ¼" plywood
6'8"

Exploded view of an insulated hollow-core door, as designed by the Shelter Institute.

sive course in energy-efficient housebuilding at the Shelter Institute in Bath, Maine. There, in a classroom setting, she learned all the basics of good construction, including the intricacies of household electrical wiring and plumbing. She and fellow students spent afternoons at building sites, gaining practical experience in framing, roofing and finishing a house, or in workshops learning special trade skills like tool sharpening and calculations of solar gain. The construction of architectural details was not neglected, either. It was at the Shelter Institute that Lesta learned how to build the door she used as a structural base for her designs.

The Shelter Institute's insulated hollow-core door is basically a "sandwich" of wood and builder's foil with a central air space inside. To build one, you make three separate frames of 1 × 4 stock, then sandwich them together with a piece of plywood and two pieces of builder's foil. The finished door is 2⅝ inches thick.

Start construction by determining the dimensions of the finished door. Measure the length and width of the door frame, then subtract ¼ inch from each dimension to allow ⅛ inch clearance on all four sides. Then cut a sheet of ⅜-inch plywood to 4 inches less than the finished door's dimensions. (The plywood helps the door hold its rectangular shape. Aspenite or Thermo-Ply are recommended substitutes because of their insulating value.)

Cut the stiles and rails of the internal frame from 1 × 4 stock. The two stiles should be as long as the plywood core, and the rails should span the distance between the stiles. Use at least three rails for good structural support. Assemble the frame with glue, nails, staples, corrugated fasteners or some combination of these. You need not join stiles and rails with mortises and tenons.

Next cut two pieces of builder's foil to the dimensions of the frame, and staple one piece to each side, dull face up. Nail the plywood to one side of the frame. This assembly is the core of the door.

Build two exterior frames to enclose the facing you'll apply to the door. Cut four stiles and four rails from 1 × 4, sizing the pieces so the frames will match the door's finished dimensions. Assemble the frames, then attach them to the core with 1½-inch #10 screws, positioning them so they are centered on the core and their edges are flush all around.

Fabricate the facing, using plywood, tongue-and-groove stock, hardwood or whatever other materials you can use creatively and attractively, and attach it to the door within the exterior frames. (The plywood side of the core will provide sound backing for any sort of facing you apply; you may need to add either more

rails or a second piece of plywood to the core to get sound backing for both facings.)

Finally, seal the channel around the edge of the door, which should measure 1⅛-inch wide and 2 inches deep. Caulk the inside of the channel to make the core air space completely airtight. Then cut wooden strips to fill the channel. Sand and stain the strips on the edges that will show, then glue and nail them in place. For extra support on the hinge side of the door, use screws rather than nails.

Once the door is hung in its jamb, tack felt weather stripping to the doorstop to prevent cold drafts.

The Shelter Institute teaches its students how to build a thermal hollow-core door, but makes few suggestions about designing the finished surface. Lesta's imaginative designs were sketched months in advance of the actual construction, in a journal she keeps for just such ideas. "There's a lot of satisfaction in expressing myself this way," she says. "I don't 'lose' ideas or designs that are important to me. I remember where I wrote them down and then go back to them. I go through two or three notebooks a year."

Keeping a notebook also allows Lesta to consider *how* she is going to work out her design ideas in the woodshop. She tackled the problem of installing the window and window frame for the front door of her studio on paper before getting out her circular saw.

Her design did require some modification of the Shelter Institute's guidelines for construction. After she cut the plywood, she cut a diamond-shaped opening in it (at what would be eye level) ½ inch narrower on all sides than the thermal glass she wanted to install. She cut the first layer of builder's foil to the same size and pattern as the plywood, and tacked it in place. Then Lesta made her 1 × 4 perimeter frame and a separate 1 × 4 window frame. She routed out a channel in the window frame to hold the glass, then beaded glazing compound into the channel and put the glass in place. She spread more glazing compound onto the first layer of builder's foil around the opening, and pressed the 1 × 4 window frame against the compound. Then she nailed the perimeter frame and the window frame to the plywood, and sealed around the perimeter with latex caulk. After cutting the second layer of builder's foil to allow for the window opening, Lesta attached the foil to the edges of the window frame and the perimeter frame to create a foil-enclosed air space. After the exterior frames and panels were attached, she added the trim.

The design of the back door of the studio, although much simpler, is no less striking. Because there is no window in it, it was not necessary for Lesta to change the basic construction in any way to suit her design. She applied cedar paneling to the exterior of the door to match the studio's cedar shingle siding. On the interior side, she made a simple picture of a sunrise in wood—a technique called marquetry.

The door's interior surface is all 5-ply, ¼-inch Baltic birch plywood—the same kind Lesta uses for her boomerangs. To make it easier to cut her design, Lesta crosscut the plywood sheet into two sections. She penciled

The front door of Lesta Bertoia's studio (exterior, opposite, and interior, right,) *is made welcoming to other artists and craftsmen by the craftsmanship that has gone into the making of it.*

her design onto the smaller section of plywood, and drilled a tiny starter hole just large enough to allow her to insert the saber saw blade; then she followed the outline of the design with the saw.

Lesta stained some of the wood to create contrasting tones for background and foreground, then glued the pieces to the inner layer of ⅜-inch plywood, fitting them back together as snugly as puzzle pieces. She stained the lower section of the birch plywood to match the foreground of her picture, glued it in place, and nailed it down around the edges.

Lesta is looking forward to the change in color that will occur as the cedar on the studio's back door weathers, but she wants to preserve the interior surfaces, so she has sealed it with three coats of polyurethane. Now, if the sunrise door gets smudged with fingerprints or crayons, Lesta can simply wipe it off with a soapy cloth.

Making hollow-core insulated doors is neither difficult nor time-consuming—if you have a work area set up for it. "I was living at my mother's house while I was making the doors," Lesta says. "My table saw and saber saw were sitting outside in *back* of the barn, so I had to run an extension cord from the house to power them. I put the doors together *inside* the barn. I had to keep moving back and forth between three places to keep the whole thing going. It took me about a week to do each door, working about two to three hours a day. I spent a lot more time on it than I would have taken if I had been really set up for it."

Although Lesta used power saws to speed up the work process, she says that a common 8- or 10-point crosscut saw could have gotten the job done just as well. She believes woodworkers should be proficient with both power and hand tools, and has shared space with other woodworkers to learn to use tools she didn't have.

Someday Lesta hopes to share her own studio space with other craftsmen. "We'd like to have some kind of art center here—some kind of place where friends feel happy to come over and work on projects of their own. We'd like to encourage people to find creativity in themselves."

VICTORIAN SCREEN DOORS

Tom Anderson is critical of the wonderfully curlicued wooden screen doors sold in an 1889 Sears catalog for $3 and $4: "Most of those doors were machine-made, and they haven't lasted. They had through mortises, and exposed end grain on the tenons, so the frames were bound to warp as they weathered." His own standards are higher. He builds wooden screen doors "to last forever"—with kiln-dried hardwoods, air-dried for an additional year in his woodshop, and strong blind-mortise and wedged-tenon joinery.

He learned about the importance of quality materials and careful craftsmanship from his stepfather, a finish carpenter and cabinetmaker. Tom began working with his stepfather when he was in sixth grade and continued helping out in the woodshop until he finished high school.

Tom had no ambitions, however, to make woodworking his career. He joined the Navy and learned other skills. When he was discharged, he married and settled down to a job as a power plant operator.

Then one day he looked at the aluminum screen door on the front of his house, all battered up and just about ready to fall off its hinges. He remembered the wooden screen doors that had opened into summer on the house he had lived in as a child in Wisconsin, and he decided to make his own "somewhat elaborate" screen door.

With that small decision, his life changed. "The screen door I made was pretty — probably the most unusual door in town," Tom explains. "My neighbor came by and asked if I would make him one, and I said, 'I'd be glad to.' Then a few weeks later, my sister came to visit from San Diego, and asked if I'd make one for her house, and I said, 'Sure.'" Soon Tom was spending all his spare time making screen doors. He decided to go into business for himself, and started his company, Creative Openings, in Bellingham, Washington.

"The year after I moved here from California was the hardest one I've ever had," he says. "Nobody in northern Washington in the middle of winter wanted a screen door for his house. It was a long, hard pull. The second year was a little better, and now, going into the third year, people are beginning to see what I do and appreciate it more."

Tom has shipped more than 200 doors all over the United States. Many of the doors have been authentic reproductions or period-style wooden screen doors for restored Victorian homes.

He will custom-design a screen door for anyone's house for a nominal fee, which is put toward the final purchase if the design is approved. He doesn't feel it is necessary to make a personal visit to the site, but requests accurate jamb measurements and several photographs of the house as a basis for design.

He likes to make his screen doors of such beautifully grained and finished woods that the homeowner isn't even *tempted* to use paint. He'll use pine where a painted screen door is what is wanted. "There are always going to be people who want nothing more than a white door to match their trim," Tom admits, "but I think you do not have to paint my screen doors white to have them look nice. I match and contrast many different hardwoods in my doors just to show a little creativity. They add so much to the house — character, uniqueness."

When he was setting up his business, Tom spent a lot of time studying wood. On a vacation trip to Africa, he stopped in at several lumber mills to see what kind of arrangements they would make to ship African mahogany, shedua, bobinga, cocobola and some of the other exotic woods he uses. Now he works with several suppliers who ship exotic and native hardwoods directly to his woodshop in Bellingham.

Part of his agreement with them is that they will tell him exactly where and when each lot of wood he buys was cut, so he can give it adequate time to season. Although all the wood is kiln-dried, wood that was logged in the mountains of Kenya requires more time to acclimate to the environment in which it will be worked than does local maple.

Tom usually uses American white ash (so stable it is standard for baseball bats), white oak (a dense, finely grained, light-colored hardwood) or African shedua (a rich brown hardwood, less expensive than teak) for the stiles, rails and bentwood laminations of his doors. He uses exotic woods (Honduras mahogany, cocobola, ebony, bacote, bobinga) and colorful native woods (osage orange and cherry) to make the hand-turned spindles and stick-and-bead forms that accent his designs.

To make a screen door, he selects seasoned, rough-sawn lumber and dimensions it. Often he will rip wide boards in half so that he can match the grain of stiles and rails. For example, if a bottom rail is 10 inches high, he will book-match 5-inch boards, and butt-join them together with pegs and waterproof glue to show off the figure of the wood.

Next Tom makes his bentwood laminations. He rips thin wooden strips—3/16, 1/4 or 3/8 inch—and cuts them to length. Then he coats both sides of each strip with glue, which helps make it pliable, and puts it on his jig— pegboard mounted on plywood, topped with movable wooden blocks that mark the shape of his curve. He adds more strips until the lamination is the proper thickness. He clamps the lamination in place and then wraps it with a rubber inner tube, which he inflates for added pressure. As each lamination dries he repeats the process to create as many curved forms as he needs. Some elaborate design elements— such as fans—must be created from several separate laminations. Since the laminations are square edged, he goes over all four edges with a router fitted with a rounding-over bit to soften them.

He also uses a router to cut blind mortises (boxwedge, which are rectangular, or foxtail, which are fan shaped) in the stiles. He cuts haunched tenons on top and bottom rails, sometimes doubling them on the bottom rail for added strength. On the lock rail, he cuts plain tenons. He splits the tenons so he can drive wedges into them, and dry-assembles the door.

When all is to his satisfaction, Tom hand-turns any spindles required by the design on the lathe, and makes simpler laminated forms and stick-and-bead forms. Then he checks them against the dry-fit door for length. Finally, he disassembles the frame, starts wedges in the tenons, and glues all joints, clamping the door so it will dry square and true.

Tom prepares the door for finishing by sanding all parts with 80-, then 120-, 220-, and occasionally 320-grit sandpaper. He finishes the door to the customer's specifications. He recommends a natural-looking finish that gives moisture protection to the wood. To accomplish this, he applies an oil-based sealer, then hand-rubs three coats of an oil-based lacquer on the door, sanding with more 320-grit between coats. He installs brass or bronze screening, and attaches brass hardware—doorknobs, latches or bullet catches, and hinge plates. The door is then ready to ship.

Tom prefers brass or bronze screening over aluminum because they are much more durable. Although a cat can claw a hole in aluminum screening or a child push through it with his hand, brass or bronze screening will

endure such abuse without tearing.

Tom ships each door with installation and maintenance instructions. Although he thinks aluminum screen door/storm door combinations are usually shoddily made, he will make a storm door insert to fit his screen doors for customers in states where the winters are severe. He also offers screen door/solid entry door combinations, and prehung screen door/solid entry door units with stained glass sidelights.

Though he is only 29, Tom Anderson certainly qualifies as an old-fashioned craftsman. "A lot of people come into my shop wanting to know how I make doors, so I show them," he says. "They tell me it's too much work, too much time.

"Well, I enjoy doing it. I enjoy doing it very much. I want the people who order a door from me to get the best-quality door they possibly can. So I lay it out carefully; I make my joints strong. I make the whole door structurally sound. And it takes more time. I think the results are worth it."

Tom Anderson takes the time to turn wood on the lathe and bend it into graceful curves to make his screen doors as beautiful as any made by craftsmen in the Victorian era.

AN UNMILLED BATTEN DOOR

"I was trained to be a mechanical engineer; also, I studied shipbuilding for two years," says Alexander Weygers. "So between that training and the approach I take to my work as an artist, I strive for a combination of aesthetics and strength. Strength, strength, I'm always after strength!"

Weygers is 80, and full of ideas, full of purpose. He carries buckets of scrap iron to his forge with an effortlessness a man half his age might envy. He teaches blacksmithing; he writes; he sculpts; he carves; he invents. He has built his house, a studio and a blacksmithing shop with his own hands, using tools he made himself.

The unmilled batten door on his studio was built without a blueprint, more by instinct than by careful calculation. After a lifetime of making and using tools, Weygers knows when he doesn't need them. With practice, he says, " . . . there are fewer and fewer instruments

A close-up of Alexander Weygers's handforged hinges shows how the salvaged metal now conforms to the shape of the wood slabs used for the door.

you reach out for, because you are utilizing your built-in instruments to the full. You learn to trust your eyes and your feel."

Weygers salvaged the lumber for his door from slabs of Monterey pine that had been dumped by local sawmills. When he realized the slabs were free for the taking, he took them. He chose slabs for walls, flooring, rafters and doors. Then, little by little, he moved hundreds of slabs to his site. His source of free lumber has since been discovered by others in the area who cut up the slabs to feed their woodstoves, but Weygers got what he wanted. He chose his slabs with bulging knots and crooks and curves—not "construction-grade lumber" at all but perfect if the builder is more concerned with the sculptural flow of forms than with conventions.

All the wood was green, so Weygers let it season in an open shed, protected from rain. Although his shed had a cement floor, he took the precaution of elevating the lumber on cinder blocks so that termites would not be tempted to nest. He used the wood as he needed it, after months—and sometimes years—of air-drying had stabilized its moisture content.

As rustic looking as the unmilled batten door is, part of its attractiveness lies in the color and smooth texture of the exposed wood. Some of the slabs shed their bark naturally, others had their bark lifted off: Weygers used his axe to make shallow cuts and then pried strips loose with his fingers.

Weygers used a door rescued from a speakeasy, covered with tar paper on one side, as the nailing surface for his battens. He chose slabs whose curves would fit together. Where they did not, he used a keyhole saw to trim them. He needed to cut only three slabs to span the width of his nailing surface and two short battens to reinforce the door at top and bottom.

To prepare the door for its hinges, Weygers clamped the slabs and battens to the speakeasy door and drilled holes through the 4-inch thickness. He also mortised the door for the hinge plate. Knowing he wanted to curve the hinges over the pine battens, he made a template by bending very stiff wire to the shape he wanted. Then he laid the door on the floor of his blacksmith shop, slabs up.

Weygers made the hinges for the door with scrap steel—a leaf spring rescued from a wrecked automobile. He describes the forging process: "All day long you reheat, reheat, reheat the steel and do your forging while the steel is malleable. Then you have perhaps a minute or two to hammer out the steel. Then, to do more, you have to reheat it again."

He worked the scrap steel into curves that followed his template. Then, while it was still yellow hot, he placed the hinge across the battens, slipped the bolts through and hammered the metal to make it follow the curves of the battens. The touch of the burning steel charred the wood, so as soon as the steel matched the shape of the slabs, he cooled the metal and stopped the charring by pouring

water over the hinge leaf. The he repeated the process to make and attach the second hinge. Finally, he propped the door on edge and bolted the whole works together.

When he hung the door in its frame, he discovered that it was so thick it would not open all the way. This did not upset him at all. He used carving gouges he had made himself to pare away some of the thickness of batten and jamb, and the door swung freely.

To latch the door, Weygers used the hardware on the speakeasy door. He added a wooden handle, which he flattened on one end and attached with screws. He explains how he found his door handle: "Whenever I cut wood for the stove, I very often see fallen limbs that are well branched, that would lend themselves to a handle. I lay them aside. I can't make myself burn them because they look so nice. Then I make use of their natural curves."

Since he lives in southern California, Weygers didn't greatly concern himself with energy-efficiency. The unmilled batten door is open so often, he didn't worry about making it airtight. If the battens shrink during the hot summer months and the tar paper shows through, he doesn't particularly care. The door completely meets his needs, so he is satisfied.

He has made only two concessions to protect the door from the elements: He gave it a coat of linseed oil to bring out the color of the wood and to make it water resistant, and he fashioned a metal awning over the door to protect it from the rain. He feels these measures are enough. Nature gives the seasoned pine a strength he can improve on only slightly. He has made these improvements, so now he goes on to other tasks.

He has this to say about the house and studio and blacksmith shop he built: "I have quite a few architects come here who fall in love with the place. One thing they always say is that they never have the amount of time they'd like to put into anything they do. When they are on

*Alexander Weygers recycled a speakeasy
door,* above, *as the nailing surface for
his unmilled batten door.*

commissions, they must always work within the
shortest time possible, so they draw blueprints.
When they look at the building after it is done
and people are using it, they wish they could
have done many things differently.

"Now I have had my time about building
this place. The whole thing grew with my living
here and applying all that I did to make ends
meet. The whole thing grew naturally. Whatever we needed, we made, and nothing beyond it
or overreaching. The natural growth of the
place—that's something that strikes many of
the young people who come here from the cities
to learn blacksmithing."

Clearly, his students learn more from
Alexander Weygers than they can ever use
at the forge.

Al Garvey

WOOD MOSAIC DOORS

Al Garvey recently got back from a trip across the Sahara. His wife, who is president of a booming pattern company called Folkwear, Inc., looked for interesting native costumes. Al sat back on his camel and watched the wind sculpting the sand dunes. Some of his impressions of that sun-baked white desert have been translated into prints; others have been made into wood mosaics on entry and passage-way doors.

Al brings art school training, hands-on experience in building and remodeling, and memories of world travel to his work as an environmental sculptor. As far as he is concerned, what he contributes is only half of what he needs to create a good design. The most important component is an intense dialogue with his client.

"I am determined," he says, "to work with my client's fantasy and not my own. I see myself as someone who can make fantasies come to life, so maximum input from my client is imperative." Al involves his clients in the design process from the very beginning. He asks them to set aside an entire day to work with him, looking through magazines for images and colors and textures that please them, and exploring memories and fantasies for visual ideas they enjoy. This dialogue sometimes turns up surprises: Although many of Al's clients focus on remembered trips to exotic places, some imagine traveling to other worlds where double moons and asteroids hang suspended in the night sky. Al enjoys the challenge of creating environments that reflect this kind of imagination.

By the end of the "design day," Al will have sketched many ideas, synthesizing as he goes along, to come up with a final working drawing which shows how he can translate the clients' fantasy into reality.

Al doesn't feel limited by having to carry out other people's ideas. "There's room for everything in art," he says. "There's room for someone else's fantasy. There's room for self-expression. But I must be allowed to make something beautiful and unique. I am not interested in placing anything more of a dubious nature on this earth, no matter how much I stand to gain personally."

Al Garvey's design philosophy points up the distinction of the craftsman as entrepreneur. He sees himself as providing a service to his clients and wants to make sure they are happy with his work. At the same time, he refuses to sacrifice his own aesthetic convictions for any price. His bottom line is not preceded by the words "net profit" or "loss," because money is not what he values most. "The most satisfying aspect of my work is, without doubt, the smiles on the faces of the people who experience it," he says.

Al designs handcrafted environments for private homes and businesses. "These environments can take the form of a single room, a suite of rooms (such as a bedroom, bathroom, dressing room), an office, a shop or an outdoor area. I design everything that goes into making up these environments down to the smallest detail, as if it is all a single piece of sculpture. If necessary, I will build, or have built, doorknobs, hinges, light fixtures, water faucets, shower heads, wash basins—in fact, anything in the environment. Of course, doors and windows are a major factor in these rooms and are designed to be an integral part of the total sculptural concept."

Al Garvey's "credentials" for such ambitious undertakings are in good order. He got his first experience woodworking in a shop class at a Chicago public high school. He enjoyed it, and set up a ShopSmith (a multipurpose stationary power tool) and worktable in the basement of his parents' house. One day, a neighbor spied him at work through the basement window and offered to help him with his projects. Since the man was a retired cabinetmaker, Al was delighted. In spite of a language barrier (the cabinetmaker was Rumanian, and could communicate with Al only in Yiddish and broken English), Al learned woodworking by watching this master craftsman's techniques, then trying them again and again himself.

Al Garvey expresses his clients' fantasies, in his own style, in every environment he designs and builds. Doors are an important part of such environments, and may be used to portray nearby mountains, above, *or the mountains and moons of another world,* opposite.

When Al graduated from high school, he went to art schools in Chicago and later in Los Angeles, where he studied painting, printmaking and design. After getting his degree, he tried unsuccessfully to make it as a commercial artist. Eventually he started making his own prints and hired himself out as a part-time carpenter to support artistic ventures.

In 1966, he went into partnership with an architect/builder doing remodeling, and gained more practical experience. "We did everything from built-in furniture to whole houses," he says. The partnership broke up after three years because the architect wanted to give up carpentry. Al was quite happy to continue working with his hands, but wanted to try incorporating more sculptural and graphic qualities into his designs. "All of my art and woodworking experience started to come together at this time to form my present vocation — building sculptural environments," he explains. "However, I still get commissions for a single piece of furniture, an entry door or a light fixture."

The double entry doors that show a blue moon caught in a tree were not commissioned as part of an environmental sculpture. In fact, the house on which they are installed was all but completed when Al was asked to design an entry "suitable for a mansion in the woods." He decided that the doors would have to be in scale with the rest of the house, and checked with the owner to see if a tree motif appealed to him.

Then, on the site, with the owner beside him, he drew up various sketches. The owner approved a design for double doors, each

96 inches high and 30 inches wide, with a wood mosaic of a tree sprawled across them. He liked the idea of the stained glass moon to lighten up the design and make the doors seem less massive. Al went back to his studio and drew up detailed working drawings. These too were approved, and he set out to build the doors.

He gathered the materials he needed. He had seasoned, kiln-dried California black walnut and koa stored under his shop. He was able to purchase the other native California and exotic woods at a local mill. He used maroc alyn (a type of bastard mahogany grown in Borneo) for stiles, rails and panels. He used butternut as the background wood in his mosaic design, and California black walnut, koa, and purple heart to sculpt the tree. For the moon, he purchased handblown stained glass (imported from Germany) in five different shades of blue from a stained glass supplier in San Rafael.

His next step was to enlarge his working drawings. He laid out 36-inch-wide brown wrapping paper on his shop floor to bring the design up to full scale. Then he divided the image into sections that could be cut from single boards. (He made these doors in 1977. If he were to do this design now, he says he would lay out the pieces "without regard for what could or could not be cut from an individual board. If a piece turned out to be too large for any board that I had I would edge-glue two or more boards together in order to accommodate it.")

He transferred his drawings, with lots of carbon paper, to a sheet of ⅛-inch Masonite. He numbered each section so he could easily reas-

semble the pattern when all the template pieces were cut out. Then, using a saber saw fitted with the thinnest hollow-ground blade he could find, he cut out the template pieces.

Next Al made the frame and panel doors that acted as a canvas for his wood mosaic. He cut stiles, rails and panel boards from maroc alyn, which is a very stable wood. He plowed a ¾ by ¼-inch groove along the inside edges of stiles and rails to receive the panels, mortised the stiles, and tenoned the rails. He built the two panels up out of edge-glued boards ¾ inch thick, and cut a ½ by ½-inch rabbet all the way around one face of each finished panel. Then Al cut holes in the panels according to his template to accommodate the stained glass. He fit the panels into the stiles and rails and glued the mortise-and-tenon joints together. Once the joints had dried, he sanded frame and panel to 120-cut. He laid the doors side by side on sawhorses with the flush (exterior) side up.

Then Al made the mosaic pieces. He laid each template on the wood (¾-inch stock) he had chosen for it and traced its outline with a ball-point pen. He cut out each mosaic piece with his saber saw and sanded to the pen line. He used a router to round off the top edges of the mosaic pieces to a ⅜-inch radius and sanded each to 220-cut.

Finally, Al assembled all the mosaic pieces on the doors, checked the fit and made final adjustments. Then he drilled screw holes through them and into the panels of the doors beneath. After attaching the mosaic pieces to the doors with 1¼-inch #10 flat-head wood screws, he covered the screws with ⅜-inch plugs of the same woods used to make the mosaic pieces.

Al turned the doors over to fit the stained glass. He caulked the glass with silicone and fit a rabbeted wooden frame, cut from the same template, over it. Then he screwed the frame into the door, and covered the screws with wooden plugs. He finished the surface of the door by sanding it with increasingly fine grades of sandpaper, all the way up to 320-grit. Then he brushed on three coats of Watco Danish Oil Finish and four coats of Varathane Danish Tung Oil Finish #67. After five years, the door has sustained water damage. This could possibly have been avoided if the finish had been reapplied every year and a half to two years.

Al has gone on to build more wood mosaic doors — as part of environmental sculptures and as individually commissioned items. He is more conscious of wood grain direction in his mosaics now and thinks in terms of high- and low-relief mosaic pieces for a stronger sculptural effect. His work continues to evolve and his craftsmanship to improve, not only because he works in his studio and woodshop six days a week, but also because he never repeats a design. "When I encounter a tough problem," he says, "I solve it. My work philosophy is simple: I sincerely believe I can build anything. I just start at the beginning and work step by step toward the end, always thinking two or three steps ahead of wherever I happen to be."

Yet he believes that his skills are not special. "Anyone can attain the needed level of skill to build wood mosaic doors like mine in a relatively short period of time. There is no mystery in the use of tools or the making of joints. To be able to make it beautiful, that is entirely another matter."

The koa tree meant something to the owners of this California house, since they had lived many happy years in Hawaii, where it flourishes. As a result, it became an important design element in the environmental sculpture Al Garvey created for their bathroom.

THE MAKING OF WINDOWS

WINDOW STRUCTURE

The word "window" came into common use sometime in seventeenth-century England, replacing a word that is more revealing of the way these openings served our ancestors. The rude openings the English built into their homes and called "windores" allowed the entry of gusting wintry air as well as the warmer, scent-laden breezes of other seasons. These openings fulfilled an important purpose—ventilation—sometimes well and sometimes at a sacrifice.

There were other factors that made windows so attractive to our ancestors that they were willing to endure attendant winter drafts. Medieval Europeans built their fortifications with splayed windows, which broadened from a narrow vertical opening on the exterior face to wide stone sills on the interior. Hidden behind such an opening, a bowman could stand and shoot at attackers from a

relatively safe and strong defensive position.

Most of the time, splayed windows served a more mundane purpose—allowing some light and fresh air into dark castle interiors. The unglazed openings helped to give castles a reputation for being cold and drafty. The beautiful tapestries woven to cover these openings have come down to us as an art form, but they are a technological achievement as well, since they are a kind of movable insulation.

When it came to glazing windows, our ancestors were inventive. The Romans used thin translucent sheets of marble and mica to cover window openings; the Japanese and Chinese used oiled paper. Even though these glazing materials allowed only dim light into the interior, glazed windows in a house marked its owners as privileged.

Modern window styles would not be so varied were it not for the invention of a process for making plate glass in the late 1600s. This technology made glass windows much more popular, but the hazards of shipping the glass by horse and carriage kept the price high. The cost continued to come down as glassmaking tools and techniques improved, but even in Colonial America and Dickens's England, the tax assessor calculated the value of a house by counting the panes of window glass.

As more people used glass in their windows, ways were found to make the windows movable so that they could be opened or closed to the outside, at the owner's preference. The French window, which became popular during the Renaissance, was rather like our

Fixed, casement and double-hung windows are favorite styles today, though they all date back centuries. Technology has made them more secure, more energy-efficient and, by providing more consistently high-quality glass, has even improved the view.

Common window styles are the double-hung, top, left; *the horizontal sliding,* bottom, left; *awning, which are hinged at the top to open outward,* top, right; *casement, which are hinged at the side, and fixed,* center, right, *and hopper, which are hinged at the bottom to open inward,* bottom, right.

panels in a large stile and rail frame. The double-hung window, which is still a modern favorite, was invented in eighteenth-century England. Innovative millworkers and joiners, and later, manufacturers, created other operable window styles — sliding, casement, awning, hopper and pivoting.

In the last hundred years, glass has become readily available in sizes and shapes to fit every application imaginable. In addition, manufacturers have developed special processes to increase its tensile strength. Although old-fashioned wired glass is still used where safety and security are concerns, tempered and laminated glass are much more popular. When heat-tempered glass is struck with a sharp object, it shatters into a thousand pieces, but the pieces do not fly. Laminated glass, made of layers of glass and plastic bonded together under high pressure, is even stronger. It is used in the front windshields of cars, in store display windows, and in detention centers to allow visibility but prevent breakage. Most building codes require the use of wired, tempered or laminated glass for windows used in and around doors to discourage forced entry. Other plastic safety glazings are often used to make skylight installations resistant to storm damage.

Of course, safety and security are not people's only concerns when it comes to window glass. As more and more people recognize the sun's energy as the earth's real income and develop building designs that collect and utilize that energy, window glass takes on new roles. Sometimes it is specially treated to absorb heat energy; at other times, it is coated to act as a heat

By using insulated glass, you can make your house more energy-efficient, whatever fuel you choose to burn. This manufactured double glazing, in which two layers of glass are sealed together with a $^3/_{16}$-inch air space in between, can significantly improve your window's thermal performance.

The technology of glass manufacture gives you more options in terms of size, shape and material properties than ever before. Knowing these options can help you decide how you want your windows to function.

A good window allows a view, permits light and air to enter at the homeowner's discretion, and is framed in a style compatible with the architecture of the rest of the house. And it achieves these goals in an energy-efficient way. In a time when all fuel costs are rising rapidly, energy-efficiency is an important design parameter—and one that is not always easy to achieve when you consider what a window is.

A window is a framed, transparent covering for a hole in the wall. If the cover for the hole isn't tight, there will be air leaks around it—infiltration. Room air, made precious in winter by the fuel and dollars you expend to warm it, will escape through cracks around loose glazing, between sash and jamb, or between casing and rough opening. Your heating system will have to work extra hard to keep your living quarters comfortable, and you will have to pay to have it working "time and a half." Infiltration can also be a problem in summer, when hot outside air flows in through any and all cracks around windows, putting an extra burden of cooling on air conditioners.

More energy-conscious housing has given windows more purpose. Above, left: In an envelope house, a window frame slatted on the interior allows warm air to flow into the living space. Above, right: An attached greenhouse collects the sun's warmth in planting beds and reradiates it into the house. Left: A skylight cut in the roof high above the living space provides soft light to the room below.

conduction—the transfer of heat energy through a material to equalize differences in temperatures on either side of the material. Glass and metal are excellent conductors of heat, and therefore make poor insulators. Wood and air are poor conductors of heat, and thus make good insulators. If you place

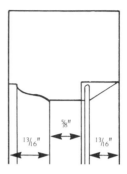

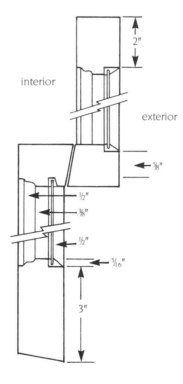

Cross section of double-hung sash, showing dimensions suggested for double-hung windows by the Architectural Woodwork Institute for 1⅜'' sash, left, and 2¼'' sash, right.

mendous amount of heat energy in winter through conduction. A single layer of glazing in a wooden frame is a better bet, but not as good as double or triple glazing in a wooden frame. With one or more insulating air spaces between layers of glazing, heating and cooling losses through conduction can be cut drastically.

Double or triple glazing can be the solution to another problem: radiant heat gain. Although glass passes visible solar radiation (light), it absorbs and checks the passage of 84 percent of the thermal energy (heat) radiated to it. The 16 percent of the thermal energy that does pass through is trapped in the interior space. Generally, this creates problems only in summer, when unshaded windows trap unwanted heat inside. Double or triple glazing obstructs this long-wave radiation, keeping room temperatures much more comfortable.

Infiltration, conduction and radiation around and through windows can cause energy losses ___ ___ for 30 to 40 percent of the heating

way to prevent such steady, sizable losses is to educate yourself about the structure of windows so you know how to make them work. Learning the terminology used to describe the parts of a window will enable you to ask your building materials supplier or glazier for what you need to make repairs or replacements or build your own window.

Glazing may be glass or acrylic plastic sheets. It is usually transparent, although stained glass used in windows may be translucent or opaque. A window with a single pane of glazing is said to have one light; a window with six individual panes of glass, set in wood or lead, has six lights.

Windows with two or more lights separated by a wooden strip or wooden grille are said to have a muntin. Window units separated by vertical dividers that are part of the frame have mullions.

A sash is the part of the window that holds the glazing. It is usually movable, but remains stationary in a fixed window. A casement window has one sash; a double-hung window has two sash. (The word is the same whether singular or plural.)

Like a door frame, a window frame has jambs and a head. All are equal in width. Jambs are vertical and the head horizontal.

The windowsill occupies the same relative position in the frame as a doorsill. A windowsill slopes downward and has a drip kerf underneath to keep rainwater from draining down the siding. Unlike a doorsill, a windowsill has no threshold. However, it's joined by a stool that projects into the interior of the house. (The stool

Again like a door frame, a window frame will have exterior and interior trim called casing, which is installed to slightly reveal the edge of the jambs and head. But don't call interior casing inside trim, because this term has its own meaning in the language of windows.

In an old-fashioned double-hung window, there are three bands of wooden strips that form the channels for the sash. The innermost one is called the inside trim or bead stop. The middle one, which keeps the sash apart, is called the parting bead or parting strip. The outermost one is the blind stop. Inside trim or bead stop, parting bead or parting strip, and blind stop all run up the jambs and across the head.

A double-hung window has two sash. The lower rail of the top sash and the upper rail of the lower sash are beveled to lock into one another. Together they form the meeting or check rail. The sash hardware that locks the windows is surface-mounted on these rails.

Older double-hung windows have the weight of each sash counterbalanced by metal sash weights; the sash and the weights are connected by sash cord. The weights are housed out of sight behind the jambs, and the cord extends over a pulley wheel set into the jamb near the head. The arrangement makes the sash opening adjustable. Modern manufactured double-hung windows often use metal spring-loaded balances or friction-mounted channels for the same purpose.

If the windows in your house were installed when it was built, they are considered prime windows. If not, you have replacement windows. Storm and screen windows are considered secondary windows.

Window quilts and inside shutters help to keep the master bedroom in this Maine house warm when the blizzards blow.

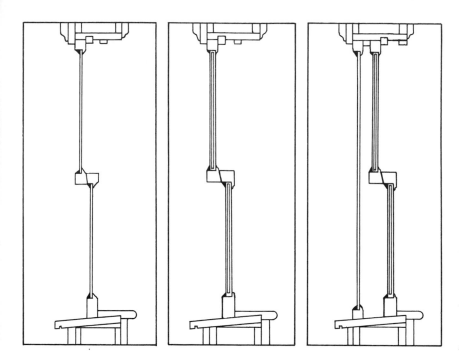

Single glazing, left, is least energy-efficient. Use insulated glass, center, or double-glaze over your single glazing, leaving a ¼'' dead air space, for an R-factor of 1.64. Increase the air space to ½'' and you'll get a value of R-1.73. Install a storm window, right, over double glazing for a value of R-2.67. Even the best glazing systems will fail if there are avenues for air infiltration between glazing and sash, sash and jambs, or window frame and finish wall.

Building materials and construction standards used in windowmaking have become fairly well established over time. The Architectural Woodwork Institute, a professional organization of woodworkers, publishes materials and construction guidelines accepted by most manufacturers and custom sashmakers. The institute recommends the use of premium or custom-grade lumber (as defined by A.W.I. standards) treated with wood preservative to prevent wood rot; it suggests that stiles and rails of sash be joined with glued mortise-and-tenon joints; and it specifies the minimum face dimensions of the sash members for adequate structural strength. It sets the face dimension of sash stiles and top rail at 2 inches and that of the bottom rail at 3 inches for a 1⅜-inch-thick sash—the most common thickness for sash that are double-hung.

However, no such dimension guidelines have been established for parts of the frame or rough opening. This may be just as well, since no two openings are precisely the same width or height or thickness. For example, the thickness of a cinder block wall faced with brick on the outside and gypsum board and plaster on the inside is not likely to be the same as that of an all-wood frame wall built with 2 × 4s between interior and exterior sheathing and siding. Since materials and methods used in construction vary, you must be prepared to make adjustments in the size of the rough opening to make preassembled window units fit.

Preassembled window units may be purchased in standard sizes and styles, directly from manufacturers or from building materials suppliers. Although small manufacturers make fewer provisions for energy-efficiency and weathertightness than the giants—like Pella and Andersen, for example—and offer much less in terms of selection, the windows they build are likely to be of good, solid construction, glazed with insulated glass, and weather-stripped around the sash.

Manufacturers, building materials suppliers and custom windowmakers use the same dimensions to describe the window assemblies they make. The width of the sash or the sash opening is equal to that of the glazing plus that of the sash stiles (usually an additional 3½ to 5 inches); the height of the sash opening is equal to the height of the glazing plus that of top and bottom rail or top, bottom and parting rail (about 6 inches).

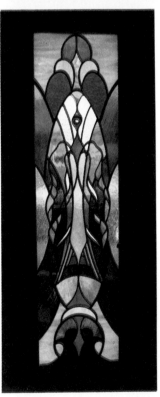

Contemporary craftsmen use stained glass to paint both representational and abstract pictures with light.

The rough opening (from doubled studs to doubled studs) is usually about 3½ inches wider and higher than the sash opening. The masonry opening formed by the outside edges of the exterior side casing, top edge of the exterior casing across the header, and the base of the sill measures approximately 1½ inches wider and 1 inch higher than the rough opening.

Instead of specifying the size of the masonry opening, manufacturers may provide you with the dimensions of the preassembled window unit. Before ordering such a unit, make sure that your rough opening is large enough to accommodate the model you want with about ½ inch allowance for expansion and contraction and squaring up.

Manufacturers make preassembled window units available in a wide range of sizes. Because good design dictates that the top of the sash openings be level with the top of the entry door, some sizes of window units are used more often than others. For example, in a kitchen where counters are set to a standard 3-foot height, a window over the sink will usually be positioned so the base of the sash opening is 42 inches above the floor and the top of the sash opening 80 inches above the floor. A dining room picture window or a double-hung window in a hall is likely to be set 30 inches above the floor to clear table or hall furniture height.

The availability of many standard sizes of glazing, consistently seasoned lumber, and precise woodworking machinery has helped to upgrade the quality and energy-efficiency of windows installed in new buildings — and so have building codes, made stricter with the energy crisis.

Other window accessories supplement existing units to add to their thermal value. Solid shutters protect glass from storm damage and burglary and add a layer of insulating wood against the cold; louvered shutters shade and give greater privacy. Screens invite the breeze — but not the bugs — inside. Storm windows add an insulating air space and cut down on winter heat losses. Awnings shade windows from the hot sun and protect the wooden frame and sash from the full force of rain. Inside, drapes and blinds provide privacy, shade out bright light, and add another insulating air space to the window area. All of these "old-fashioned" window treatments are being rethought and sometimes redesigned. The modern-day window quilt, with its batting of fiberfill and reflective vapor barrier, held snug against the window casing, is akin to those medieval tapestries — but uses the technology of the Space Age rather than that of the fourteenth century.

The state-of-the-art technology manufacturers make available to builders and homeowners alike is impressive to any serious window shopper. Sash and frame come as a unit in casement, bay, bow, awning, mullion, horizontal sliding, double-hung, hopper and stationary picture window styles. Prefinished window components are available to build windows to non-standard sizes or shapes. Insulated glass is standard in several manufacturers' lines Storm and screen inserts may be purchased for the double-hung window, a very popular style; triple-glazing inserts are stocked for casement, awning, double-hung and horizontal sliding windows.

If you spend your money wisely, you can get a very carefully thought out, energy-efficient window that is nearly maintenance free once installed. Sash and frame parts will have a wooden core for insulation value. The wood will be treated before it is machined to resist fungus and rot. Sash and frame parts may be coated with rigid vinyl sheathing that eliminates the need for homeowners to repaint and repair weathered sash and frames; it also saves the builder the labor and materials costs of painting window trim. Both rigid and foam weather stripping are used to seal the movable parts of sash and frame.

Top-of-the-line windows are expensive, but in time the initial investment is likely to be recouped in money saved on fuel bills. Unfortunately, replacement parts are costly and sometimes difficult to get. Like new cars or women's fashions, manufactured windows are being "improved" all the time. By the time the crank on your casement window gives out, the part may have "improved" so much that those kept in stock won't fit your window unit.

Architects, energy consultants and businessmen have been inventing solutions to window problems with the same technology used by the corporate giants. For example, many custom storm and screen sashmakers use extruded plastic to mold sash, frames and triple-track storm and screen inserts, because it keeps out cold better than aluminum. Once installed, the storm and screen window units can be left in

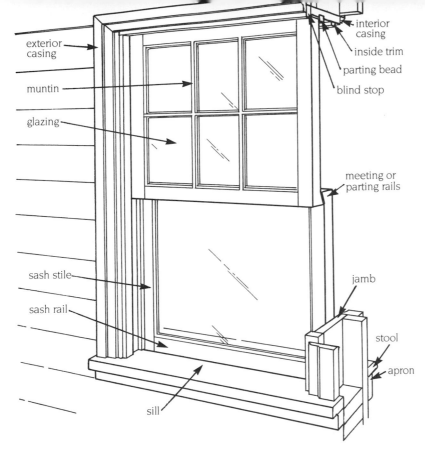

exterior casing

muntin

glazing

interior casing

inside trim

parting bead

blind stop

meeting or parting rails

sash stile

sash rail

jamb

stool

apron

sill

The parts of a double-hung window.

place year-round, and the inserts can be removed for easy cleaning.

Why make your own windows when manufactured windows have so much going for them? If you have more time and patience than money, you have a good reason to make your own windows. If you want windows made from especially durable or attractive materials — like oak or cherry — you have another valid reason. If a window style not commonly stocked, like triple casement, appeals to you, you have strong motivation.

The craftsman can vouch for others: the pleasure of working with your hands, with good tools and good materials; the satisfaction of completing something that will serve its purpose well in the home; the sharing of the value of craftsmanship with those who appreciate it.

Working with art glass to create a pleasing picture or pattern allows you to explore your ideas of beauty. Working with ordinary glass to build an energy-efficient window demands design discipline and an understanding of the relationship of form and function.

To meet the challenge of handcrafting a window that combines these values, you must learn the skills of the carpenter and glass artisan, and strive for the thoughtfulness of the architect and the attention to detail of the craftsman.

Sound ambitious? You may have already discovered that you are competent as a handyman as well as an accountant, or that you are a green thumb gardener as well as a good cook. Discover some more of your talents and aptitudes. Try making a window. It can open up a new view of yourself and your world.

MAKING YOUR OWN WINDOW

Good window design requires careful consideration of the climate of the area in which you live. Fixed windows can work well in an air-conditioned desert home or, when double or triple glazed, in a cabin in Alaska. The same window style, installed in a Pennsylvania farmhouse, may be a bad design choice because it doesn't allow the homeowner to take advantage of gentle spring and summer breezes for natural ventilation. Being acquainted with the climate of the area in which you live, you are in the best position to decide how many layers of glazing you need and whether ventilation is more important than having an airtight window.

A window may seem to be a very static design element in a house. In reality, however, it is dynamic—a meeting point and passageway between your house's internal climate and that outside. You can work with the natural climate or make your house a refuge

from it. In general, it costs less in terms of energy and money to design with nature, more to design in opposition to it. Technology can be used to support either choice.

Think about the architectural style of your house as well as the climate. A simple double-hung window looks inappropriate on a Colonial—but add a wooden grille to create the effect of multiple lights and suddenly the design feels right. The double-hung replacement windows in a Victorian row house may look a little "off," too—until you add a muntin bar. If you have a very modern house, you may want to use only casement and sliding or fixed picture windows. The simple lines and uninterrupted surface areas of these styles make open, uncluttered living space seem even more light and airy. Whichever style you choose, remember that the tops of the sash openings on the ground floor must be even with the top of the entry door, and on other floors even with all other sash openings on the same floor. If not, the sense of design balance will be thrown off.

When you have decided what window style best suits your wants and needs, sketch it on paper. Show how you want it to look from inside the house as well as from the outside. Draw it in cross section. Mark the dimensions of each member of the window sash and frame.

When ordering lumber, look for quality. You want your window to *last* and not to warp and leak air. Pine is a favorite of many sashmakers. Cedar, cypress, Douglas fir and redwood may also be used with satisfying results. A hardwood such as oak is a good choice for durability. You can make your selection by asking your

Windows need not be regular rectangles to function well and conform to the design of the house in which they are installed.

building materials supplier which woods are available locally and which are least expensive. In most cases, these will be the same, since locally milled lumber requires fewer middlemen to bring the wood to you. If you plan to make several sash and frames at one time, take this into account when ordering lumber.

Assemble your tools before beginning the project. If you work only with hand tools, you will need backsaws and handsaws, hammer, planes, drill and chisels. Power hand tools that might be substituted are the circular saw, router, and drill. If you have a more complete workshop, you'll be able to do a lot with the radial arm saw or table saw. Glazing the windows will require a putty knife.

If you intend to make sash for several openings, you might make an old-fashioned carpenter's tool—a "story pole." Use a long strip of lumber to mark off the top and bottom of your rough opening, measuring from the floor. If you are making and installing windows in new construction, the tops of the rough openings for them should be level with the tops of the rough openings for entry doors. When you have finished installing your first window, you can mark other vertical reference points on the story pole to make your other installations consistent with the first. If you use the story pole for openings of different sizes, mark reference points for each size in a different color.

Rough-frame a window much as you would rough-frame a door. Cut away wall studs to create an opening. If both sides of the window frame will align with the wall studs, simply reinforce the studs from floor to ceiling and install doubled studs for rough sill and header. If the window is not on center, make the rough sill and header long enough to span the opening between two wall studs, and then install a board called a trimmer between rough sill and header to mark the side of the actual rough opening. Shortened studs called cripples can also be spaced at intervals under the rough sill to offer extra support when the rough opening is especially wide.

As you work, make sure that the header and rough sill are level and that trimmers, studs and cripples are plumb. The finished rough opening should be square and true, and its dimensions adequate to accommodate the sash and frame, with ½ inch clearance.

Fixed windows, though the simplest form of window, can still be a challenge to make. Follow that timeless bit of carpentry advice: Measure twice, cut once. Use a framing square to insure that boards are warp free and the rough opening, frame and sash are square and true. To ease installation, do as much of the assembly and finishing work as you can on the horizontal surface of your worktable. This may save you some unnecessary and dangerous work on a ladder.

Make the sash, the part of the window that holds the glazing, rather like a frame and panel door. Cut tenons on the rails. Mortise the stiles. Rabbet all four pieces along the inside edge of the exterior faces to receive glazing. Thinly coat mortises and tenons with waterproof glue, then assemble the pieces by hand. Clamp the pieces together, checking for squareness as you work. Let the glue dry.

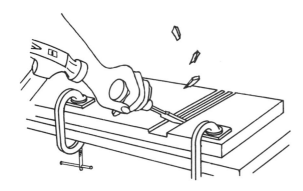

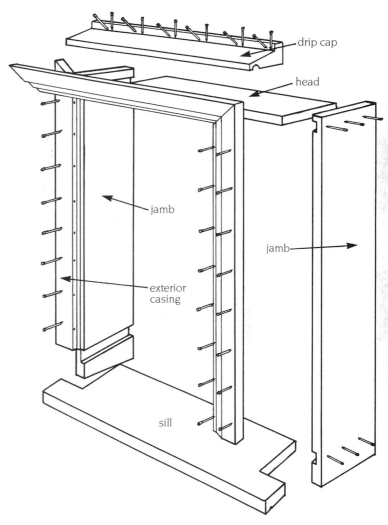

Then assemble the window frame for the sash on your worktable so that sash and frame can be installed in the rough opening as a unit, rather than piece by piece.

In frame construction, the head and side jambs must span the thickness of the wall; that is, the width of the stud plus the finished interior wallboard and exterior sheathing and siding. Cut the side jambs of ⁵⁄₄ stock to fit vertically between header and rough sill. On the inside face of each side jamb, cut a dado to receive the head. Each should be about one-third the depth of the side jamb, and wide enough to accommodate ⁵⁄₄ stock.

You can cut dadoes with a table saw, radial arm saw or router. If you are using hand tools, make several cuts with a back saw across the area you want to dado out and then go after the waste wood with a chisel.

Cut the head from ⁵⁄₄ stock, too. It should span the sash opening and extend into the dadoes on the side jambs.

Prepare the side jambs to receive a sill by cutting dadoes on the inside faces, angled to slope about 15 degrees downward to the outside. You can figure out where the top edge of the dado should be for fixed, simple casement and double-hung windows this way: Lay the side jambs on your worktable with the exterior edge of each jamb down. Fit the head in place, then lay the sash in position. Pencil-mark the inside face of each side jamb at the bottom edge of the sash, allowing ¼ inch for clearance. Using a try square, extend the mark across the jamb. Measure along this line and mark where the exterior face of the sash will line up.

Above, left: *Making a dado on a jamb using hand tools.*

Above, right: *The parts of a window frame.*

119

To mark the actual sill line, set a sliding T-bevel to a l05-degree angle using a protractor. Position the bevel against the jamb with the blade intersecting the point marked on the horizontal line. Scribe a line along the blade.

The sill can be fabricated of ⁵⁄₄ stock or of something thicker. In any case, let the thickness of the stock dictate the width of the dado for the sill. The dado's depth should equal one-third the thickness of the jambs. Cut the dado.

Fabricate the sill next. It must be cut to extend the full width of the window with the exterior casing in place, then notched to fit in between the jambs. The lip of the sill, when in place, must extend about 2 inches beyond the exterior edge of the jambs. Since the sill slopes 15 degrees, the inside edge and the edge that fits against the exterior sheathing at the base of the casing must be cut at the same angle. Finally, under the lip of the sill, you need a drip kerf.

Assemble the frame by fitting the parts together by hand. Check for squareness, then nail through the side jambs into the head jambs and through the side jambs into the sill to make the frame structurally stable.

Cut a stool to fit over the sill, wide enough to project 2 inches into the room and long enough to fit around the jambs on both sides to the width of the inside casing. Use a table saw to cut a rabbet on the underside of the stool at a 15-degree angle to accommodate the sill. If you don't want to tackle the angled cuts with your power tools, you may be able to buy precut sills and stools at your local building materials supplier.

Put the stool in place against the sill, but don't nail it down. Now fit your sash. Plane a bevel on the bottom rail of the sash to fit the slope of the sill. When it fits comfortably within the frame with the bottom rail butted up against the stool, mark the placement of outside and inside stops on jambs and head. Cut the stops, either molding or square-edged wooden strips, to fit. Nail the outside stops in place. Remove the stool from the frame; it will be attached

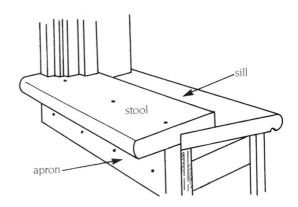

Cross section of a windowsill.

along with the interior trim after the window is installed.

With the frame assembled, you can now cut and attach the exterior casing. Proper fit is important, so the pieces should be cut to fit and nailed to the frame one at a time. Start with one side piece. Cut a 15-degree bevel across one end of a length of the chosen stock. Fit the bevel against the sill and mark the casing for cutting at the head. Remember that you'll want to maintain a reveal of about ¼ inch. Square-cut or miter the casing, depending upon the style of installation you prefer, then nail the piece in place with finishing nails. Cut and install the other side casing in like fashion.

A square-cut head casing is easy to cut and install, but getting the proper fit when mitered joints are used takes some know-how. Miter one end of the casing stock. Lay it atop the side casings upside down, with the point of the head casing miter lined up with the point of one of the side casing miters. Mark the head casing where the other side casing miter touches it, then cut it to that mark. The piece should drop right into place. Nail it fast.

Finally, when all your sash, frame and casing pieces are cut and assembled, prepare the wood surfaces and paint them with primer or wood preservative to protect the wood. If you intend to stain the wood, you may find it convenient to apply at least the first coat now.

Next glaze the sash. Cut the glass ⅛ inch narrower and shorter than the dimensions of the rabbeted sash opening. Work a bead of glazing compound into the rabbet, then gently press the glass pane in place, push glazing points

in place, and smooth glazing compound around the perimeter of the glass with a putty knife to seal it to the sash. Take care that the glazing compound is not visible from the other side of the sash above the rabbet.

Install the glazed sash in the frame. Lay the frame, with the exterior casing attached, down on your worktable. Fit the sash in place against the outside stops. Secure it in the frame with the inside stops. Attach the stops with screws to allow you to remove the sash easily should the wood or glass ever need repair.

The moment to install your window unit has now arrived. Place the unit in the rough opening from the outside. If the unit is large or is to be installed on the second story or higher, have a helper hold it in place while you go inside and check to see that it is level and plumb in the rough opening. Insert wood shingles as needed between side jambs and studs, head and header, and sill and rough sill to shim it square and true. Pack loose fiberglass insulation in the openings between shims to protect against infiltration.

Secure the unit by driving finishing nails through the exterior casing into the doubled studs. Since finishing nails have almost no head, drive the first nail only partway into the side exterior casing and then check with a level to make certain the unit is still square and true. Sink the nail. Drive the second nail partway into the exterior casing on the opposite side of the window, and check again with the level before sinking it. Once the unit is anchored, simply nail at regular intervals down the jambs and across the head to secure it. Make or buy a drip cap (a piece of casing that is sloped and routed

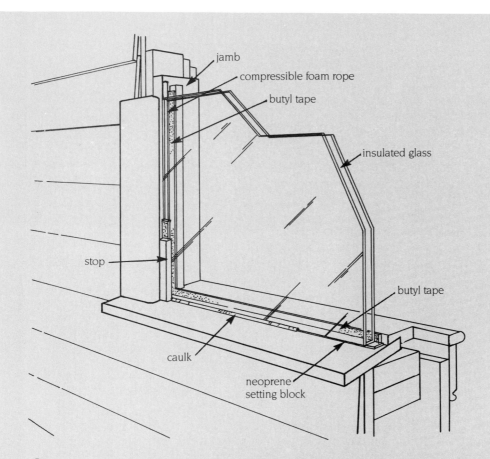

jamb

compressible foam rope

butyl tape

insulated glass

butyl tape

stop

caulk

neoprene
setting block

An Energy-Efficient Fixed Window

Because a fixed window is a permanent covering over an opening in a building, it is possible to make it quite energy-efficient by choosing your materials wisely and installing them so they seal up all possible avenues of air infiltration.

For the highest insulating value, double-glaze with clear acrylic plastic sheets or glass, leaving a ¾-inch air space between the two layers. Insulated glass (two layers of glass sealed together with a 3/16-inch air space in between) is a good second choice.

Rather than making a separate sash to hold your glazing, make it integral to the frame. Line the frame with an inside window stop, then seal the stop to the frame and create a bed for the glazing with a good glazing compound or caulk. Rest the bottom of the glazing on a neo-prene strip. Install the center stop or "spacer" that goes between the layers of glazing, and put the exterior glazing against it.

If you use insulated glass instead of making your own double glazing, apply weather-stripping tape against the exterior face of the inside window stop, and compressible foam rope along the joints. Snap a couple of neoprene setting blocks along the edges of the insulated glass along all four sides to fit it snugly against the frame.

Cut outside stops and line the interior face of the stops with weather-stripping tape. Nail them into the jambs and header. Caulk across the bottom of the glazing. Attach an aluminum drip cap along the header. Finally, apply finish trim.

on the underside with a drip kerf), and attach it to the wall sheathing directly above the top exterior casing. Set all the nails and putty the nail holes. Caulk all around the exterior casing to seal the window frame to the house.

Finally, cut and install the interior casing. Fit the stool over the sill to butt up against the sash. Toenail it into the sill with finishing nails. Cut an apron, a square-edged piece of trim as long as the stool and wide enough to reach from the base of the stool to the drywall-covered studding of the rough opening. Center the apron below the stool. Nail through the apron, the interior wall sheathing and into the rough sill and cripples to secure it. Add side and top interior casing, setting it back $1/8$ to $1/4$ inch to form a reveal. Cut the interior casing to fit, just as you did the exterior casing. Use finishing nails to nail the casing to the jambs and through the interior wall into the doubled studs of the rough frame. Set all nails and putty the nail holes. Caulk the joints of the interior casing with the wall to lessen the avenues of unwanted air infiltration.

Paint, stain or varnish sash, frame and casing inside and outside to seal it for weather and warp protection.

With this style of fixed window, you can double-glaze by using insulated glass in the sash or by adding a storm window insert flush against the inside or outside stops during the cooler seasons. Do take some precautions to prevent condensation problems between the layers of glazing. Weather-strip your fixed window well if the storm sash fits against the outside stop. Vent moisture from between a fixed window and a storm sash against an inside stop by drilling a tiny weep hole into the insulating air space from the underside of the sill.

Casement windows must be made by someone with sophisticated carpentry skills and a penchant for problem-solving. If they open inward (as they usually do in England), they are hard to seal against leaks. If they open outward (as they generally do in the United States), they are difficult to screen because the operating hardware must extend through the bottom rail of the casement sash and sometimes the screen as well to be accessible. Manufacturers make such hardware, but the necessary pieces aren't generally available, though they can be specially ordered.

With a little ingenuity, however, you can make a simple outward-swinging casement window—one you can screen—but you will have to make some of the hardware, as well as the sash and frame, yourself. Fortunately, you need not be a metalsmith to make this hardware. Woodworking skills will do.

You can, and probably should, prefabricate the window unit in the shop, although you'll have to hold off on finally installing the stops, stool and interior casing until the unit is installed. In brief, you build the sash and the frame, hang the sash, fabricate the window stay and lower stop, prime the unit, glaze the sash, then install the unit.

Start with the sash. Build it just as you would for the fixed window detailed in the preceding pages.

The frame, too, is made like that for the fixed window. Cut the head, jambs and sill, and dado the jambs. Assemble the pieces together without nailing, then cut and dry-fit the stool.

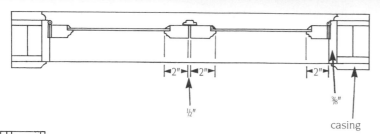

Far left: *Cross section, side view, of an outward-swinging casement window.*

Near left: *Cross section, top view, of double casement windows with flat astragal meeting stile.*

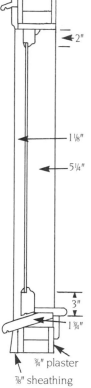

If necessary, lightly tack the frame together, so you can fit the sash in place.

The sash in the casement window cannot fit as tightly as the sash in the fixed window. It is in effect a door, and it must have clearance between it and the frame, just as a door does. The sash must be beveled along both the bottom rail, to fit the sloping sill, and along the lock stile, to permit it to swing open. Plane the bottom rail first, then when the fit is correct, plow a drip kerf down the center of the rail. Don't plane the lock stile until after you've hung the sash.

Mark a line on the sill where the sash's drip kerf will align. Remove the sash from the frame. Knock the frame apart and plow a kerf along the line. Then, using a hinge-mortising bit in a router, excavate about ⅛ inch of material from the kerf to the sill's outer edge. This recess, in combination with the sash's drip kerf, will promote drainage away from the sash's edges, which are not protected from the weather by exterior stops.

Reassemble the window frame and nail it together. Be sure it is square and true.

Cut and attach the exterior casing.

Normally, inside window stops line only the joint of the sash and the jambs and head. To make the casement window operable, however, you must make an additional stop that rests on the stool and is mortised for a movable window stay. It must be high enough to accommodate the stay and allow it a little play. With this in mind, cut the inside stops to fit. Nail the head stop in place, but only tack the others in, since they rest on the stool, which you will have to remove to install the window unit.

You are now ready to install the sash. Fit the sash against the stops, making sure the stiles are plumb and the top rail level. Use cardboard shims to hold the sash in place while you mark hinge placement. You should have about 1/16 inch clearance all around between sash and frame; if the unit will be stained rather than painted, a slightly tighter fit is okay.

Hold a set of hinges in place and scribe a line along its top edge on both jamb and sash. Remember to position the hinges above and below the rails, as appropriate. Work carefully to align the hinges accurately. Mark the placement of the second set of hinges in the same way.

Remove the sash from the frame and clamp it in a vise with the edge of the hinge stile up. Extend each of your marks across the edge using a square. Set a leaf of the hinge at one mark and cut into the wood around it with a utility knife. Move the leaf to the other mark and repeat. Use a chisel to pare out the wood within the outlines. Work very carefully, since there's no completely satisfactory way to build up a mortise that's been cut too deeply. Both pairs of hinges *should* sit flush with the edge of the stile. Keep paring until they do, then attach the hinges to the sash.

At this point, you may want to recheck the marks on the jambs by refitting the sash (with the leaves of the hinges attached to the sash open). Pencil-mark correct placement. Remove the sash. Then score the outlines of the hinge leaves marked on the jambs with the utility knife. Pare out the mortises.

Temporarily hang the sash—drive a screw or two in each hinge—so you can see how much the lock stile must be beveled to allow the

124

window to close. Remove the sash and plane the bevel. This again is a trial-and-error procedure; proper fit is important.

Once the sash is hung, you can install the casement fastener and its strike. And you can fabricate and fit your window stay, which you will need to operate the window. Cut a strip of sturdy wood (like oak) ½ inch thick, 1 inch wide, and long enough to extend from the midpoint of the bottom rail of the casement sash to the inner edge of the stool when the casement is wide open (at a right angle to the building). Drill $\frac{5}{16}$-inch holes through the strip of wood at regular intervals.

Next attach your handmade window stay to the sash. Turn a screw eye into the bottom rail at the center, about ⅜ inch above the level of the stool. Turn a hook into one end of the stay, then hook it over the eye.

Another trial-and-error task—the mortising of the bottom inside stop—follows. First you must remove the stop from the frame. Then drill a ¼-inch hole in the center of the stool and glue a bit of ¼-inch dowel in it, so that it projects about ¼ to ⅜ inch above the stool. The stay fits over this peg and each hole in the stay represents a different open position.

Open and close the window, marking on the stool the limits of the stay's action. Transfer the marks to the stop, then mortise the stop—a notch will do and is easier to cut—to accommodate the stay throughout its action as the window is opened and closed. The fit should be as snug as possible (without causing binding), so start small and slowly enlarge the mortise or notch until it is just the right size. Tack the stop in place, then install a second peg in the stool to

A simple casement window.

casement operator

crank handle

Manufactured casement window hardware.

125

Since this house was built by Lothrop Merry around the time of the Revolutionary War, many have lived here and admired the view of the Atlantic framed by these windows.

hold the stay as far out of the way as possible when the window is closed.

Paint sash, frame and all other wooden window parts with primer or wood preservative. When they are dry, glaze the sash just as you would for a fixed window.

Install this casement unit as you would a fixed window. Fit the unit in the rough opening from the outside, carefully shimming it so that it is level and true. Nail through the exterior casing and the wall sheathing into the studs to secure the unit to the house. Add a drip cap. Caulk around the casing to prevent unwelcome drafts from getting in. Inside, install the stool, apron and inside stops. Add the interior trim and caulk around it.

Paint or finish the window as you prefer. Press an adhesive-backed foam weather strip along the inside stops lining the jambs and header, and along the stool. In winter, this will help make your window secure against invading cold. In summer, you can clip a screen insert within the opening created by the inside stops and still work your window stay. A few bugs will probably find their way through the stay's mortise, but not many. And you will be able to enjoy the fresh air that motivated you to build a casement window in the first place.

This simple casement window may not offer as much security as some would like. Since the hinge pins are exposed to the outside, make sure that they are nonremovable.

If you would prefer instead to use modern casement window hardware, make sure that you can get all the necessary parts. Modern casement hardware consists of a casement fastener (latch and keeper or strike), tracks, casement window operator and hinges.

126

When the crank handle is turned, the gear mechanism hidden inside the casement window operator is engaged. As its wheels turn, the arm extends, sliding along a metal track set in a rabbet on the lower face of the bottom rail, and the window opens. When the crank handle is turned in the other direction, the arm draws in and the window closes.

The casement fastener and operator are handed. If you plan to make a casement window that swings outward and to the right, buy right-hand hardware, and vice versa.

The casement window hinges are silent partners to all this mechanical activity. Each hinge looks and works like a small metal elbow: A flat metal bar is secured to the frame at two points. The point farthest from the hinge jamb acts as a pivot for another metal bar, which is anchored to the sash. When the window is closed, these two bars sit on top of one another. When the window is open, the two bars, connected at the pivot, angle apart so that the sash remains supported.

Unfortunately, the hinges must be specially ordered. Make sure your building materials supplier has a source for them before you try to make a modern casement window.

Making this style of casement window will probably be a challenge. The sill must be rabbeted to accommodate the hinges without interfering with the operator. The head must also be rabbeted for the hinges. The sash must be rabbeted at top and bottom for the tracks in which the operator arm slides. The sash must be mortised for the arm and gear mechanism. And because manufacturers dis-

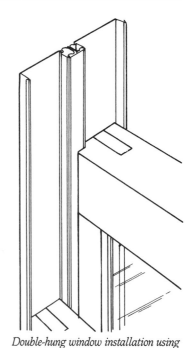

Weather-Stripping Double-Hung Windows

With your double-hung windows sliding up and down easily in pressure-mounted channels instead of operated by sash weights and pulleys, there are far fewer avenues in and around the jambs for air to infiltrate. You can boost the energy-efficiency of your window still more by using vinyl tubular or spring metal weather-stripping along the joint of top rail and head, bottom rail and sill, and parting rails. Vinyl deteriorates when painted, however, so take this into account when deciding which weather stripping is for you.

tubular vinyl weather stripping

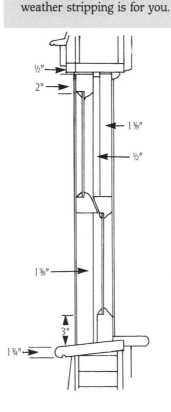

Double-hung window installation using pressure-mounted window channels.

127

The double-hung window becomes a design element with impact when care is taken to tie it to its surroundings with appropriate trim.

tribute casement fasteners, operators, crank handles and tracks as replacement hardware, the instructions on the packages don't detail how to prepare frame and sash. The size of these parts varies according to model and manufacturer. If you are determined to build a modern casement window, you are left with two choices: work from the hardware to calculate the dimensions and placement of mortises and rabbets, or use an existing casement window (fitted with hardware like that available to you) as a pattern.

Double-hung windows used to be considerably more complicated to make than the fixed and casement windows just described. In older houses you can sometimes still see double-hung windows that have metal weights hidden behind the jambs to counterbalance the weight of the sash. The frames incorporated pockets for these weights, which made them difficult to construct. The jambs were mortised for the sash-cord pulleys and grooved for the parting strip, which divided the space between the inside and outside stops into two channels, one for each of the two sash to ride in. Access to the weight pockets was through the jambs, to allow broken sash cords to be replaced.

Modern technology has made the construction and installation of double-hung windows an old-time carpenter's dream. With the use of pressure-mounted window channels, the carpenter can say goodbye to sash weights, pulleys, ropes, parting strips, stops and the weight-pocket behind the jambs.

The starting point for such a construction project is a pair of replacement window channels of the sort made by Quaker City Manufacturing Company of Sharon Hill, PA 19079. The channels are used primarily to renovate old windows whose sash rattle in their wooden channels. You rip out the old stops and parting strips, fit the sash between the new metal channels and slip the pieces into the old frame. The channels are then fastened in place with nails or screws.

You don't have to be renovating a window, however, to make use of the channels. Begin making your double-hung window by cutting and assembling the parts of the frame and its exterior casing, just as if you were making a fixed or casement window. Next cut two pieces of the window channel (using a hacksaw) to fit between head and sill.

Constructing the sash is the hardest part. They must be constructed largely of 1⅜-inch stock so they fit into the window channels, but the bottom rail of the upper sash and the top rail of the lower sash (which form the meeting or check rail when the window is closed) must be made of thicker stock. This allows them to accommodate the parting strip, which keeps the two sash about ¼ inch apart, so that they can move freely up and down in the channels.

The upshot of this is that you'll have to buy material for these sash that is custom-milled to the necessary thicknesses. In addition, you'll have to reverse the normal sashmaking procedure. Instead of cutting tenons on the rails and mortising the stiles, do just the opposite. This will allow you to use the thicker stock for the bottom rail of the upper sash and the top rail of the lower sash, and to bevel them to get the fit you want at the meeting rail. Once both sash are assembled, cut rabbets at both ends of the check rail to accommodate the parting strip (on

This homemade skylight brightens up a dimly lit interior, cheering up people and helping plants to flourish.

129

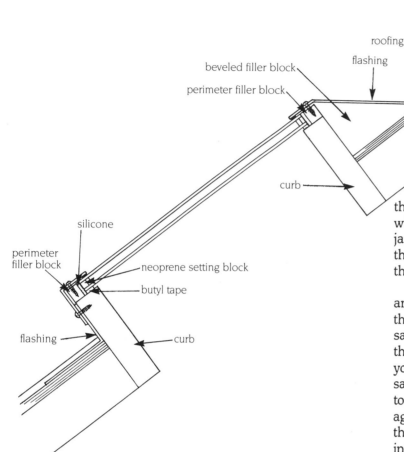

roofing

flashing

beveled filler block

perimeter filler block

Cross section of a double-glazed skylight.

curb

silicone

perimeter
filler block

neoprene setting block

butyl tape

flashing

curb

the bottom rail of the upper sash, these rabbets will be along the interior face and toward the jamb; on the upper rail of the lower sash, they will be along the exterior face and toward the jamb).

After the sash are assembled and the sash and frame are primed, glaze the sash. Then fit the sash into the window channels—the upper sash in the outside tracks and the lower sash in the inside tracks. With the frame horizontal on your worktable, exterior casing face up, place sash and channels in the frame opening. Check to see that the window channels are plumb against the jamb. If not, straighten them. Secure the channels to the jambs with screws at regular intervals. Run a bead of caulk along the joint of each channel and jamb, then cover the joint with an outside window stop.

Install your double-hung sash unit from the exterior of the house. Shim it as necessary to make it sit square and true, and then secure it with finishing nails through the exterior casing. From the room side, pack the opening between jambs and doubled studs with loose fiberglass insulation. Attach the stool, apron and interior casing. Paint, stain or varnish to finish the installation. Add storm or screen window inserts to make living with the seasons a pleasure.

Skylights, whether purchased or hand-made, have become increasingly popular with homeowners who enjoy the idea of living with nature. When carefully located, constructed and installed, skylights can brighten up dark rooms, allow cooling breezes to enter the house, and even contribute to the workings of a successful passive solar heating system. When poorly made or carelessly installed, however, skylights can give you good practice for being a volunteer fireman—when it rains heavily, you'll find you are pulling duty in the bucket brigade! So don't consider making your own skylight unless you are confident of your carpentry skills, curious enough to do thorough research, and capable of patience with your tools and materials and with yourself and your mistakes.

Unfortunately, space does not allow a thorough treatment of skylights in this book. If you are interested in making and installing your own skylight, you will find some of the books listed in the appendices helpful in getting you started. Please see in particular *The Skylight Book* by Al Burns and *Solarizing Your Present Home*, edited by Joe Carter. *Solarizing* contains a helpful chapter on the construction and installation of both manufactured and owner-built skylights called "Project: Skylights," written by the book's assistant editor, John Blackford. Doing a little of this "armchair carpentry" can make all the difference between a skylight that is a pleasure to have in your house and one that's a pain in the neck.

If you have never tried woodworking before, start with a simple project. Don't be afraid to make mistakes, but learn from them when you do make them. Build and install a fixed window before you attempt a casement. As you gain in experience, your skills will improve, and your confidence will increase.

If you are already proficient with woodworking tools, then take on whatever task will challenge you. If you work with patience and care, you'll find you can build whatever you want—

Window walls allow these homeowners to enjoy an unmatched view of the San Francisco Bay.

WORKING WITH GLASS

The stained glass shop by night is full of beacons that borrow light from the interior to send a message of warmth and welcome to those passing by. Many are drawn inside knowing they may have a moment's pleasure without paying a penny for the color, the dance of light and line. The forms of beauty are there for even the most casual browser to enjoy. Those who come with more serious intent—to buy or to learn the craft—do so having recognized that the sight of something beautiful can lift our spirits all the days of our lives.

There are many ways of working with glass to create beauty. Bevelers grind and polish glass until it scatters light into rainbows like a prism. Etchers use stencils to frost it with representational or abstract forms. Engravers dance a diamond-tip across it to do delicate line-drawings. Stained glass artisans use transparent and

opaque colored glass in combination with ordinary window glass to paint with light. Some of the most beautiful window glass designs combine ways of working with glass. A stained glass window showing birds by the water, for example, may be etched and engraved to highlight the birds' feathers and the froth of the waves lapping the shore.

Today's glass is the product of a long technological evolution that has greatly refined the manufacturing process and increased the number of ways in which the quality of glass can be controlled. In its molten state, glass can be drawn into thin fibers, blown by hand to form goblets and by machine to form beakers, ladled into molds, and poured flat for windows. Its color can be changed by the addition of metal oxides; its texture can be varied by running the molten glass through rollers that make it look rippled or hammered; its hardness, its ease of cutting, and its tendency to shatter or simply fracture on impact can be controlled by the rate at which it is heated and cooled, or annealed.

To make glass, silica, soda and lime are heated until molten and then annealed to form a solid in such a way that no crystallization occurs. It is this lack of long-range ordering of the atoms that makes glass rigid, yet easy to break, and responsive to changes in temperature and the pull of gravity. The oldest window in England, found in a house on Bradford Street in Braintree, Essex, is thickest at the bottom of the pane, due to the slow descent of

Glass is enduring. Much of the handblown window glass installed in European taverns and hostelries during the Middle Ages is still in place. A glassblower made broad glass by cutting the ends off a handblown cylinder, slitting its length, and then returning it to the ovens where it would flatten into a roughly rectangular shape. He made crown glass by blowing a large sphere, flattening it and attaching it to a pontil opposite the blowpipe mouth, and then freeing it from the blowpipe. He widened the hole made by the blowpipe with a wooden paddle until the glass flattened out into a disc. Once annealed, cut to fit, and hung in a window opening, this glass showed a distinctive "bull's-eye" or raised circle of glass that marked the point where it was supported by the pontil.

Perhaps nothing in the 4,000-year-old technology of glassmaking so revolutionized its production as Frenchman Bernard Perrot's discovery of a process for making sheet glass. In *Journey to Paris*, published in 1698, Dr. Martin Lister, physician to Queen Anne, included an account of a visit to a Paris glassmaking factory that used Perrot's technique:

"The Glass-House out of the Gate of St. Antoine well deserves seeing ... For I saw here one Looking-glass foiled and finished, 88 inches long and 48 inches broad; and yet but one quarter of an inch thick. This, I think, could never be effected by the Blast of any Man ..."

The specific details of the process invented

Above: *A round window with clear glass frames the view from a thirteenth-century castle in Beja, Portugal.*
Right: *A detail of a window in a door on a houseboat shows a special way of working with glass: etching clear and irradiated glass for a design that glows with subtle color.*

but it is thought that rolling the molten glass to flatten it was an essential step. Up until a few years ago, glass manufacturers used the same process to make modern sheet glass. After heating silica, soda and lime to 2,800°F, the glassmakers let the molten glass flow out of the mixing tank in a continuous ribbon. This ribbon passed between a series of steel rollers, which helped to flatten and gradually cool it. Then the glass was cut and packaged as sheet glass or polished, ground and cut for sale as plate glass. The polishing and grinding steps made the surfaces of the glass absolutely parallel, so that there was no distortion.

Today's glass manufacturers eliminate these steps by using a process developed in England in the 1960s. When floated over molten tin, glass becomes absolutely flat and free of imperfections before it moves along the assembly line on rollers, cooling gradually on its way to be cut. The U.S. government has approved the use of the terms "float" and "plate" glass interchangeably, since the quality is equally good. Not only do manufacturers use float glass in standard window sash, but bevelers, etchers and engravers apply their tools and talents to it to create freestanding sculpture, wall hangings and special windows.

Glass bevelers run pieces of float glass under a series of grinding wheels to make a rough bevel progressively smoother. Then they use polishing wheels to hone it to prismlike perfection. Straight-line beveling is done entirely

guided by hand. If you are tempted to take up glass beveling, test the seriousness of your interest before buying equipment. Even a small set of grinding and polishing equipment costs over $1,000. However, some craft supply shops carry manufacturers' lines of precut beveled pieces that are usually quite sufficient for most amateurs' and craftsmen's needs.

Glass etchers use hydrofluoric acid to create opaque white silhouettes. Since even the acid's fumes are caustic, it is extremely dangerous to use or even to store in a home workshop. As an alternative to acid, it is possible to do wheel etching, as well as engraving, with a small power tool called a glass sculpture router. Beveling, etching and engraving remain less popular than stained glass work, perhaps because the delicate effects they make possible are so much less dramatic—and less visible from a distance— than those obtainable with colored glass.

Glass engravers begin to learn their craft using the same kind of diamond-tipped, high-speed electric drill rented out by police stations for homeowners to use to inscribe identification on possessions. This inexpensive tool, which is little bigger than a ballpoint pen, is available at many hardware stores and at all lapidary shops. It can be fitted with tiny burrs that allow you to sand, grind and polish small areas of glass.

Begin your experiments in glass engraving using a scrap of glass clamped to the end of a worktable. Make sketches until you come up with a design you like. Use a sheet of carbon paper to transfer your design to the glass. Then,

Clear glass, beveled along its edges to refract light like a prism, or etched or sandblasted to show patterns of white on white, can have as strong an impact as sculpture or painting.

135

with your whirling drill point, follow the lines on the glass. Remember that it may take a little time for you to become comfortable with the feel of the power tool in your hand, and to hold it steady as it moves over the glass. Do yourself a favor by keeping your initial designs simple.

A glassworker's studio, like a woodworker's shop, can be simple or elaborate. Along with hand tools like glass cutters and pliers, the glassworker may have a power trim saw (like a table saw), band saw, router, drill, grinder, belt sander and polish arbor. If he specializes in etching glass, he may have sandblasting equipment; if he paints on glass, he may have a high-heat gas or electric kiln. The stained glass artisan is likely to have the simplest and least expensive setup of all—another reason for the continuing popularity of this medium.

Stained glass artisans need a specially prepared work surface. Make your own using ½- or ¾-inch particle board cut both longer and wider than the dimensions of the wall hanging or window you intend to make. Tack molding along two edges of the particle board to form a corner guide. Lay this on your worktable.

Stained glass is no longer just for churches. Now it is inexpensive enough to use in the sitting room in the window behind the woodstove, or in the skylight over the kitchen counters.

Gather some of your tools from around the house—ruler, scissors, hammer, household oil, X-Acto knife or razor blade, sponge, small paintbrush, stiff wire brush, linseed oil and putty. Make some of your tools. For example, two single-edge razor blades shimmed and taped together along the dull edge can substitute for the double-bladed pattern shears used by professionals to cut patterns for lead came. You are likely to find all the rest of the tools you need at a stained glass supply shop. Buy an 80-200 watt

soldering iron with a ⅜-inch-diameter tip, 60/40 solder, a lead vise for stretching lead came, a came knife, a stone and file to keep the came knife sharp, and a lathekin for opening channels and smoothing joints. You will also need several lengths of lead came with a U-shaped channel to finish off the perimeter of the stained glass assembly you make. Purchase lead came with an H-shaped channel to join all the glass pattern pieces within the assembly. Of course, you will also need glass.

If you visit a stained glass supply shop with the intention of buying glass but with no particular design in mind, you may come away so overwhelmed by the choices available that you buy nothing. Alternatively, you may buy colors and textures of glass that you must later set aside because they do not work in your final design. To avoid these frustrations, make your first visit to the stained glass supply purely an exploratory mission.

Ask to see both antique and rolled glass. Antique glass, which is handblown and flattened like broad glass, has imperfections that add to its beauty. It may be solid (a single clear color), seedy (full of tiny bubbles), streaky (swirled with different colors), reamy (swirled with different textures), crackled like chips of ice or flashed so that only one side is colored. All kinds of antique glass are quite transparent, but differ in the ease with which they may be cut. Solid antique glass is available in an enormous range of colors and hues. With other kinds of antique glass, the color choice is more limited.

Rolled glass, made by machine, is also called cathedral glass. It may be clear or opales-

Stained glass artisans' designs may be representational, show a stylized motif, or be abstract. Lead lines are as important a design element as color, and add structural strength. Brace bars reinforce larger designs, like the lightwell shown at bottom.

(Continued on page 140)

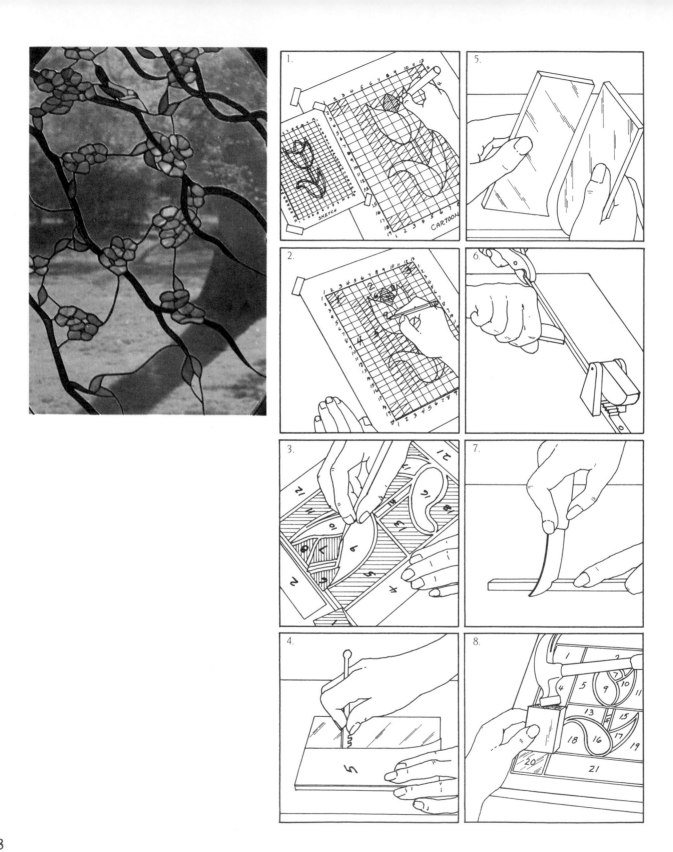

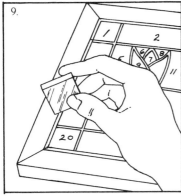

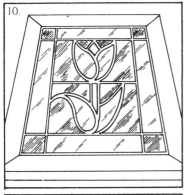

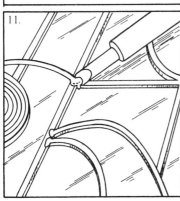

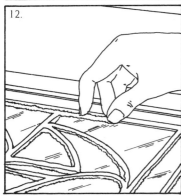

Basic steps in stained glass windowmaking are as follows:

1. Make sketches on tissue paper until you have a design and color combination that satisfies you. Use a grid system to enlarge your sketch to a full-scale "cartoon," drawing in lead lines to the true thickness of the came and coloring as you did your final sketch.

2. Layer cardboard and carbon paper under your cartoon. Tack securely in place and retrace your design, drawing both edges of the lead came and numbering the pattern pieces for the glass in a sensible order.

3. Take the cardboard and cut your pattern pieces for the glass using an X-Acto knife and shimmed razor blades. When all the pieces are cut, reassemble in order to check fit and completeness.

4. Use your cardboard patterns to cut pieces of glass in the colors you have chosen. Fix the pattern to the glass with sticky-faced masking tape, and follow its outline to make your score.

    Pull the glass cutter steadily toward you, beginning and stopping the score ⅛'' from the edge. Never go over a score line twice! Dip the cutter in oil between scores.

5. Using the other end of the cutter, lightly tap the underside of the glass along the score line. Snap the glass by holding it firmly in both hands and exerting upward pressure under the score line. Assemble the pieces on the cartoon to make them easier to locate in the sequence you need them.

6. With one end of the H-channel lead came in the vise and the other held firmly by pliers, stretch the came, straightening it as necessary and opening the channel with a lathekin.

7. Cut two lengths of came an inch longer than the length and width of your window design. Use a came knife, cutting with a gentle rocking motion against the soft metal.

8. With the cartoon taped in place on your working surface, position the two lengths of came against the molding strips and tap in the corner piece of glass. If its edges do not fit within the lead lines on the layout, score the glass and "groze" it, chewing away the excess glass with the teeth of the glass cutter.

9. Use the glass to measure the lengths of lead came that will hold the corner piece of glass in place within the design. Cut the came 1/16'' shorter than measured to allow for solder. Proceed sequentially to fit glass and lead came together, grozing when necessary, until you finish the pattern.

10. Finish the came border by trimming the two border pieces already in place to their proper length, and fitting the remaining two to frame the design. Tap two more molding strips in place on your working surface to secure the window for soldering, with all its corners square.

11. With a small paintbrush, dab oleic acid flux on all the joints to be soldered. Uncoil the roll of solder so the end lies flat over a joint, and touch it with the hot soldering iron for a few seconds. Seal all the joints this way, using about ⅛'' or ¼'' of solder.

12. Cement the lead came to the glass on both sides with metal sash putty. Wipe away excess with an orange stick. Clean with sawdust and a bristle brush, then ordinary window cleaner.

    Before carrying out a stained glass window design, consult the sourcebooks listed in the appendices for more detailed instructions.

Salvaged Glass

With 12 years of experience in stained glass behind her, Rhonda Dixon doesn't find working with three-dimensional glass objects a problem. The window above contains some scraps she has salvaged and recycled—Depression glass plates, broken canning jars, crystal stemware, lots of bottle necks, and "cullet"—waste glass discarded by the glass-blowers in Silver Dollar City, Arkansas. All of the small glass pieces were sealed together with copper foil work, the larger ones leaded with standard H-channel lead came.

Rhonda uses the copper foil technique whenever she must join pieces of glass that are small or in an intricate pattern. She wraps copper foil around the edges of each piece of glass, assembles all the pieces according to the pattern, and secures them in place on her work surface with molding strips and tacks. Then she brushes powdered flux all over the copper foil lines, and solders them along the entire length (rather than just at the joints) to create a lead bead. When she is finished, she turns the panel over and repeats the fluxing and soldering process.

Most people wouldn't want a composition of broken glassware hanging in their living room window, but in Rhonda's greenhouse it serves an important function. When the summer sun pushes the inside temperature up to 120°F, a soft breeze blows in through all the bottle necks, helping to cool the plants and contributing to cross-ventilation. When the temperature begins to drop in the fall, Rhonda simply corks all the bottle necks—a novel but effective way to keep out drafts!

cent and textured on one or both sides. Solid rolled glass is generally not quite as transparent as solid antique—the colors tend to be murkier. Opalescent glass may be translucent, semiopaque, or opaque. Ease of cutting varies. See what you can learn about the glass by holding a sheet up to the light and by running your fingers over its surface. The stained glass supply store owner can apprise you of other differences.

Once you know the material choices available to you, you can plan an intelligent window design. For your first project, keep your design small—under 8 square feet—and use simple shapes. (As you improve in technical skill, you can tackle larger and more complicated projects.)

Even a simple stained glass window can have a stunning effect if it suits the context and color scheme of the room in which it will be installed. Your choice of subject matter is important but it is the creation of pleasing relationships of form (or shape), texture and line that will make your design successful.

In working with stained glass, you are painting with light. The colors you choose have an emotional impact. Reds and yellows are warm, vibrant, exciting; blues and greens cool and restful. Color intensity makes certain elements of the design seem to come forward and others seem to recede. For example, the red of a tulip's petals and the emerald green of its leaves appear to be in the foreground when placed in a design "field" of palest blue. You will learn a great deal about light and color by working with the medium itself.

Painting on stained glass can add detail, intensify color.

As you advance in technical skill, you may be tempted to try painting on glass. There are special paints made just for this purpose, composed of ground-up glass particles and metallic oxides. These vitreous paints are the same kind used on the most enduring and beautiful painted stained glass windows of the Middle Ages. They are made long-lasting by kiln-firing, which fuses the painted colors to the glass surface.

Building codes allow leaded art glass to be used as glazing in and around entry doors. Standard float glass is generally used in all other windows in a dwelling. If you learn to work with glass, you need not be bound by this convention. Your windows can be bordered with bevels or etched glass lacework, or colored to catch the light of the sun. Those who visit your house will feel your special joy in it, and be glad of their welcome.

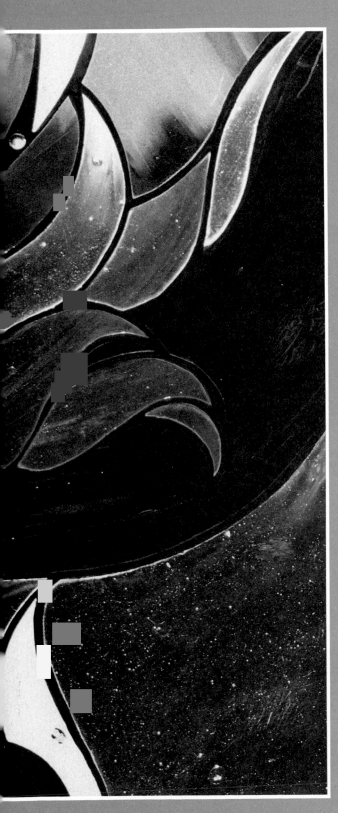

A GALLERY OF WINDOWS

Larry Golden

FUN FRAME-UPS

When Larry Golden answers his phone, he quips, "Stormy weather" or "Tundra region"—whatever pops into his head. His window designs are just as unpredictable and fun. He likes to take advantage of salvageable or inexpensive materials on site—and to follow his whims.

The diamond window, shown at left, brightens the inside of his own house. "This design evolved from the materials at hand," he says. "I had a lot of old lights from a barn I had torn down, and I had some leftover Plexiglas from some skylights. I was curious about doing my own thermal panes, and I wanted a diamond window, a traditional shape in the barns in the Midwest."

In building his house, he left a 54 by 54-inch opening for a window. Once he decided on the shape, he put in the rough framing. Since the walls are framed with 2 × 6 studs and covered with ¾-inch pine

144

sheathing on the outside and ½-inch sheetrock on the inside, he ripped 2 × 8s to the finished wall thickness, and used them to make the rough framework.

Then he made his sash, stiles, rails and muntins of 2 × 6 pine. An ardent believer in glues and nails in place of complicated joinery, Larry confesses, "I just got crazed and dadoed out everything so it would all lock together. It'll *never* fall apart!" He assembled the sash, glued and clamped it, then installed it in the rough-framed opening. When it was dry, he cut outside stops from 1-inch stock and tacked them in place. He beaded silicone caulk up against the stops and pressed in Plexiglas, precut to fit the sash openings. Next he cut and nailed 1 × 4 strips to act as spacers between the layers of glazing. To finish his sashmaking, he again positioned the old barn lights against the spacers and installed the inside stops.

Larry used pine for the diamond windows, and coated it with Pentothal to protect it from wood rot. He prefers the natural color of weathered wood to any other finish, so he did not use paint or stain.

He feels that he overbuilt the diamond window sash, and would simply install the glass and stops in the rough framing if he were to build another like it. He's also learned since building the window that more is not necessarily better when it comes to creating an insulating air space between layers of double glazing. (According to architect William Langdon, author of the book, *Movable Insulation*, the R-value of an air space between layers of double glazing "increases up to about ¾ inch," at which point air convection currents in this space begin to offset thermal gain.)

Overall, Larry regards the window as a success. "After seven years," he says, "I still enjoy the design. When the sun hits the window, the panes of old glass ripple in the light. And in the winter, it is pleasant to look out and see the river." The window functions well — letting in light and warmth and showing off a special view like a fine picture frame.

One of the secrets of Larry's success in designing and building windows of all different shapes is that, whenever possible, he uses acrylic plastic — Plexiglas is the most familiar brand — as his glazing material. He is enthusiastic about its possibilities.

"Plexiglas comes in all sorts of shapes and sizes. It's available in sheets, rods, blocks — and you can get it in a rainbow of colors. You could use it in place of stained glass."

Larry buys the Plexiglas he uses for windows in 4 by 8-foot sheets. (Locate the supplier nearest you by looking under "Plastics" in the Yellow Pages.) Plexiglas comes with a paper coating on both sides to protect its surface from scratching. According to Larry, there are other advantages to this paper coating. "You can draw the outline of your design right on the paper and use the pattern as a cutting guide. The paper prevents the Plexiglas from chipping or splitting along the edge." He adds this caution: "Exposure to moisture or direct sunlight over time will make it difficult to pull the paper away from the Plexiglas. Once this happens, you can get it off with a cloth soaked in denatured alcohol, but it's a pain. So it's best to cut the plastic and pull off the paper immediately."

Acrylic plastic will shatter under the sharp blow of a hammer and chisel and scratch at the touch of a nail, but you can cut it with ordinary

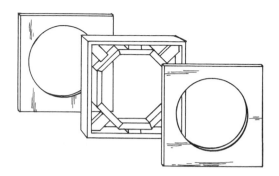

Larry Golden makes a plywood and board lumber sandwich to rough-frame a round window.

woodworking saws. Use a belt sander to smooth any rough edges. Make holes for nails or screws with an electric drill. When working with acrylic plastic, use power tools to get smooth, chip-free edges — and *keep the paper coating in place until all cutting, sanding and drilling* are done. Take the paper off only when you are ready to put the glazing against the window stops.

In recent years, plastic manufacturers have worked most of the kinks out of acrylic plastic sheet glazing. It no longer yellows with exposure to the sun. Sometimes condensation forms between the two layers when the glazing is first installed, but Larry says the condensation usually dissipates about a year after installation. He continues to use it happily — because it allows him to push the possible shapes of windows to the limit.

Of course, oddly shaped windows require special framing. When Larry wants to frame and hang round sash, he asks the builder to leave him a square opening large enough to accommodate the dimensions of his window with a few inches to spare. Then he makes a "sandwich" of plywood and framing lumber with a hole cut out in the center to receive the sash. For example, if a house is framed with 2 × 4s and Larry wants to install a 3-foot-diameter window, he asks the builder to leave him a 4-foot-square opening. He cuts a sheet of ½-inch plywood sheathing in half, clamps the two 4 by 4-foot sections together and cuts 3-foot-diameter openings for the round sash in the center of the plywood with a band saw or jigsaw. Next he makes a square frame of 2 × 4s to the dimensions of his opening and nails one piece of

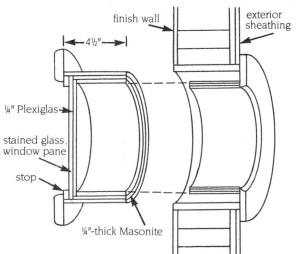

finish wall

exterior sheathing

4½"

¼" Plexiglas

stained glass window pane

stop

¼"-thick Masonite

By using Plexiglas in place of glass, Larry Golden can cut his glazing with woodworking tools to whatever shape suits his design. He uses interior and exterior window casing with just as much imagination to emphasize the window form.

Larry Golden's method of framing and installing a round sash.

The zigzag windows, double-glazed with Plexiglas and caulked for weathertightness, succeed in achieving energy-efficiency without sacrificing unique design and personal style.

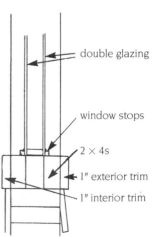

double glazing

window stops

2 × 4s

1" exterior trim

1" interior trim

Cross section of the double-glazed sash and frame of the zigzag window.

plywood to it. He turns it over, adds blocking around the hole and from the hole back to the square frame, then nails on the other plywood piece. Finally, he fits it into the square opening the builder left for him.

To make double-glazed round sash fit his rough frame, Larry laminates two separate frames of Masonite that slide together like cake tins. He nails the exterior trim into the outside edge of the smaller frame. He seals Thermopane against its inside edge. Then he fits a center stop and the interior glazing into the rabbeted inside edge of the larger frame. He seals the glazing with silicone caulk and nails the inside window trim in place to act as a stop. Such a neat installation is proof that there's no *one* way to frame a window.

The zigzag window is certainly a fantasy made real with wood and Plexiglas. "This one really just happened," Larry says. "I met this fellow who was building a house, and he wanted to put something pleasant and exciting into it. He asked me to do a wall and a window and a door, something very dynamic.

"So I went over to the site one day and framed the zigzag window. I had asked the framers to leave me a 5-foot rough opening, and I just filled it in. I looked at the opening, and *then* I did some doodles on scrap lumber, decided where it was going, and I was off!" Since the house was framed with 2 × 6s, Larry cut out three matching 2 × 4s for each section of the curved sash frame. He used a saber saw to cut the irregular shapes. Then he nailed the frame in place so that all the 2 × 4s were flush.

Once the sash frame was anchored in the wall opening, he had a friend hold his sheets of Plexiglas against the openings so that he could pencil-mark his cutting patterns. He clamped two thicknesses of Plexiglas together and cut straight sides with a circular saw, curves with a saber saw.

Then he installed the glazing in the openings. First he attached the exterior stops. Then he beaded silicone around the joint of exterior stops and frame, and pushed in the Plexiglas. He cut ⅜-inch cedar plywood strips to act as center stops between the layers of glazing and nailed them down flush against the exterior Plexiglas. He caulked again, fitted the inner glazing to the openings, and attached the inside stops. Then he nailed on trim fabricated from 1-inch stock.

"For conceiving a design on the spot," Larry says, "I feel that this was one of my best efforts. The owner is happy with the window, which makes me feel good." Like most craftsmen, Larry is satisfied only if his custom-made windows look good *and* perform well.

Larry's freewheeling work style did not develop overnight. He studied design and sculpture at art school and continues to work as a sculptor. He feels fortunate to have learned construction techniques from old-time carpenters. Both the art training and years of experience in construction contribute to the design and function of the doors and windows he makes now.

He is a craftsman—but not a traditional one. "I am of the 'make-do' school. I will use whatever is available/handy/practical. I do not mind traveling to do my work—or, if someone wants me to, I can make a window or door and ship it to them. So long as the homeowner wants something exciting and fun, that is what counts for me."

Bruce Fink

POLYESTER RESIN DOME SKYLIGHT

Bruce Fink was already an award-winning industrial designer when he graduated from the University of Illinois in 1962. When it came time to look for work, he decided not to seek a position in a giant corporation. Instead, he set up his own woodworking and metalworking shop and began bidding for commissions. Almost 20 years later, he is still happily self-employed, designing doors, windows and whole houses that end up on the pages of glossy magazines.

He's not concerned with such publicity, though. Simple word-of-mouth advertisement has gotten him commissions as diverse as making lost wax and cast bronze metal doors for the Chicago and Los Angeles Playboy Clubs and building a 12-foot circular polyester resin skylight for his country neighbors, the Smiths.

Bruce's design process is a little unconventional. He sits down in a favorite chair, closes his eyes, and visualizes the object he wants to design as if it were drawn on a blackboard. When he wants to see a cross section, he simply changes the image on his mental blackboard. To check for structural strength and determine joinery, he rotates the object, inch by inch, 360 degrees. If he feels he needs to view the object from other angles, he does so.

Most designers spend many hours at a drafting table to come up with a final blueprint for a design; those who can afford it use computer graphics to shortcut the process. Bruce's "headwork" allows him to save time, money and paper. If he draws a blueprint, it is simply to show the final design to the client for approval.

Bruce's satisfied customers of two decades will testify that his designs *work*, but the method he uses is not for everyone. He is extremely inventive and imaginative — a gifted sculptor, an award-winning designer, and a careful craftsman. Those of us whose talents lie in other directions would be wise to use paper and pencil, all the geometry we know, and any drafting equipment we can beg or borrow to come up with a complete set of working drawings on which to base construction.

Bruce, who is very sensitive to environmental factors that affect design and living conditions, takes into account thermal values, air and moisture infiltration, and durability of materials and their response to ultraviolet radiation when designing a window.

The Smiths' skylight is double glazed, well sealed, and has a drip kerf. In addition, a special compound was added to the clear plastic resin mix used to make the skylight panes to prevent them from yellowing and aging in the sun.

Bruce suggests that anyone making a skylight for the first time start small. "Working on a window of this size certainly points out the absurd amount of extra care and work needed to do such a simple task as turning the window over. Had the frame been 6 feet in diameter (rather than 12), I could have handled it alone, without the use of an overhead chain hoist. If I had another man's help, we could have easily turned over a circular window up to 8 feet in diameter."

The process Bruce used to build the Smiths' skylight can be summarized as follows: He built the frame, and used it to make a plaster mold. He made a second mold of opaque plastic. He built up a layer of clear glazing on the exterior and another on the interior of the opaque plastic mold. When these layers had cured, he separated them from the mold. He assembled copper latticework inside the frame to help support the glazing, and then installed the skylight. However, building and installing the skylight is no weekend project. He estimates the project took him 400 hours from conception to finished installation.

He built the circular frame of fir 2 × 8s and plywood. To cut his 2 × 8s to the needed curves, he set up a special cutting jig on his table saw. He made a rigid A-frame of 1 × 2s, and attached it to his working surface so that it could pivot freely. He nailed through the arms of the A-frame to hold his 2 × 8s in place. Then he simply swung the 2 × 8s in an arc over the table saw to cut his curves, raising the saw blade ¼ inch with each pass until the 2 × 8s were cut through. This gave him a smooth-sawn surface,

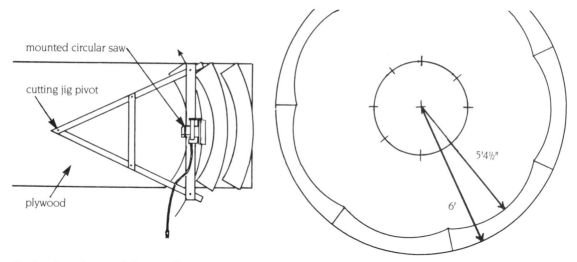

mounted circular saw

cutting jig pivot

plywood

5'4½"

6'

eliminating the need for sanding.

He reassembled the 2 × 8s front to back, with glue and 3-inch Sheetrock screws (driven in with a power hand drill). He cut the ends so that they would be in line with a radius from the center of the circle. Then he laminated a double thickness of the 2 × 8s, staggering the joints, until he had a round wooden form with a 12-foot diameter.

Next he modified his jig to cut plywood. He clamped the plywood over a working surface, and set the pivot of his A-frame jig into it in such a way that its arc would define a circle with a radius of 80 inches. Then, where he had nailed the 2 × 8 to pass it over the table saw, he mounted a small circular saw to make his cut. To cut the inside curve, he moved the pivot of the A-frame and widened its angle to define a circle with a radius of 42 inches. Then he moved the cutting jig again to cut more curves.

He screwed a double thickness of the plywood arcs to the wooden form to complete the frame, and waxed the screw heads so that they could be easily removed later on. Then he cut down 20-foot lengths of ⅜-inch rigid copper pipe to make the latticework (visible from underneath the finished skylight). He flattened the ends, drilled screw holes in them, and began weaving the pipe into a lacy pattern. This was not as difficult to do as it sounds—the pipe walls were thin enough that the pipe could be bent by hand.

He wove the copper pipe into a seven-pointed star. He used a fine copper wire (stripped out of some electric cord) to wire it securely. First he wired together the three pipe ends that fell at each point of the star, then he twisted short lengths of copper wire at the 14 inside junctions. "At this point," he says, "the entire latticework is evident. If any changes are to be made, this is the last real moment to make them."

When he was satisfied, he screwed the copper latticework to the wooden frame, and greased it thoroughly along all lengths and at all intersections so that the plaster that would be applied to it next would not bond to it. Then he set it on three sawhorses, convex side down, and tested the strength of his chain hoist to lift the frame, and his own strength to turn it over.

Bruce prepared to build his plaster mold by spreading a drop cloth under the frame to protect the floor. Then he mixed up batches of #1 molding plaster (he used 200 pounds altogether) in 5-gallon plastic buckets. He dipped pieces of burlap (cut 6 inches larger than each separate opening in the copper lattice) in the plaster, and draped each cloth so that the edges hung over the copper rods by several inches to counterweight the sagging cloth in the middle. He draped the center hole first, and then every other opening. Once the plaster-soaked cloths had hardened, he trimmed their edges. Then he wet down the hardened cloths and proceeded to fill in all the remaining openings with more

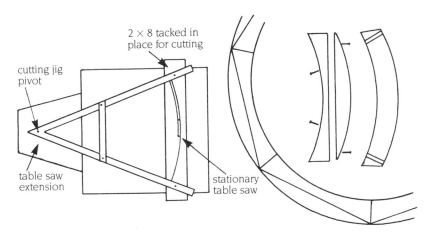

2 × 8 tacked in place for cutting

cutting jig pivot

table saw extension

stationary table saw

Opposite: *Cutting jig for plywood and method of reassembling plywood for making frame.*

Left: *Top view of cutting jig for 2 × 8 and method of reassembling cut 2 × 8 to form an arc of the circular frame.*

Below: *Side view of copper latticework screwed to frame.*

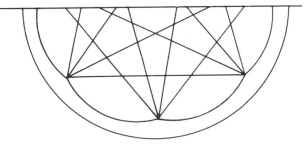

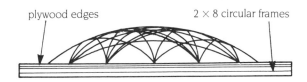

plywood edges

2 × 8 circular frames

plaster-soaked cloths. When these had set, he brushed on a final coat to smooth the inside surface. Where it was necessary to push the plaster into small cracks, he used a butter knife.

When the inside of the plaster mold had dried, Bruce turned the unit over and smoothed the other side with brushed-on plaster. He trimmed back hardened plaster around the copper rods so that half of each rod's circumference was exposed. Since this surface provided the shape for the inside of the finished skylight, care had to be taken to make it uniform.

"If you were only interested in a single-glazed skylight," Bruce says, "this plaster could be made thicker, treated and sealed and used as a final mold. However, the mold would probably be somewhat damaged after one cast so that you could not use it again. Since I wanted to double-glaze the Smiths' skylight for a higher insulating value, I made a second mold from my plaster mold that could withstand repeated castings."

Bruce prepared the plaster mold to receive a fiberglass-resin covering by melting 8 pounds of dark brown microcrystalline petroleum wax (sold by most oil companies for foundry use), and brushing it over both sides of the plaster mold. To make it really adhere, he torched it so the wax would sink into the plaster.

He applied the polyester resin in several coats. For the first coat, he mixed 1 gallon of Allied Resin Corporation's Contact Molding

Resin 33-031 with 2 ounces of a catalyst (MEK Peroxide 60 in a ½ to 2 percent solution). He added 2 cups of Cab-O-Sil (a fine-particle silicon dioxide used to thicken resin mixtures) and enough molding plaster powder to make it pastelike. Then he applied the mixture to the concave surface with a brush.

He let the first coat harden for about half a day, and brushed on a second, thicker coat of the same mixture, being careful to fill in any cavities around the copper rods. This coat hardened, and he applied a touch-up coat, as needed, to level out the surface.

Next he mixed up resin and catalyst alone

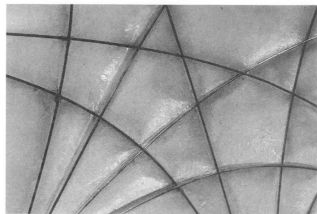

in 1-gallon batches and brushed it on the hardened surface. Then he laid small pieces of fiberglass matting onto this still-wet surface, embedding it in the resin/catalyst mixture. He overlapped the pieces of matting for structural strength. Over the course of several days, he covered the entire mold surface with four to six layers of the fiberglass matting and resin/catalyst mixture. Between coats, he checked the surface for improperly raised fibers, and sanded as necessary with a coarse-grit paper on a disk sander.

To finish the mold, he applied another thick coat of the first mixture and then a thinned version of the same mixture to smooth the contours and give gloss and hardness.

Bruce separated the finished fiberglass-resin mold from its plaster base by removing all the screws in the plywood. He disassembled the copper latticework, cleaning each pipe and numbering it so the latticework could be reassembled later and used in the final clear cast skylight. Then he drove wooden wedges between the 2 × 8s and the plywood, and lifted the unit up with his chain hoist so he could work underneath it, chipping out any remaining plaster.

When he had cleared away the plaster mold, he went over both sides of the fiberglass-resin mold inch by inch, checking for imperfections. When he was satisfied that there were no undercuts, he applied three coats of Allied's Mold Release Clear Paste Wax and a top coat of Plastilease 512B (also by Allied) to both sides of the mold to prevent any bonding between the surface of the mold and the skylight glazing.

He mixed Allied's Clear Casting Resin 32-033 with 1 ounce of MEK Peroxide 60 (in a 3 percent solution) in ½-gallon batches. He added a small amount of Tinuvin 328 (an ultraviolet absorber) by Ciba-Geigy to further increase the resistance of the finished clear cast skylight to yellowing on exposure to the sun.

He brushed this mixture on in thin coats, embedding and overlapping small pieces of the fiberglass matting to give it structural strength. After the first coat had set, he put on another coat, and gradually built up the surface until he had six coats. Then he turned the mold over and repeated the application process on the other side.

He advises, ''The resin surface remains tacky when first hardened. All successive coats should be applied during this period for best adhesion. Brush the final coat on without laying on any fiberglass matting so that the resin fills in slight fiber indentations and gives a hard, glossy surface.''

Bruce reassembled his latticework and reattached it to the plywood of the frame. He sprinkled brass dust powder over the convex surface of the skylight as an additional ultraviolet inhibitor. (Bronze powder, marketed by Leo Uhlfelder Co., in New York City, is sold in many art supply stores as #22 Finest French Gold Leaf.) Then he allowed both translucent lay-up panes to cure for several weeks at room temperature.

When both interior and exterior glazing layers were no longer tacky to the touch, he separated them from the mold by wedging the edges apart and hosing water in between so they

floated apart. With the help of a crew, he hoisted the interior glazing (sealed to the frame, braced with the copper latticework) up to the top of the Smiths' tower. He nailed the frame into the beams of the tower walls, sealed them together with caulk, and added a drip kerf. Then, again with the help of a crew, he lifted the exterior glazing to the tower top, sealed it to the interior with caulk, and made a second drip kerf.

Bruce has two final suggestions for anyone who wants to tackle this big project: Clean your brushes and your bucket between coats or you will have a very difficult time salvaging your tools, and protect your hands by wearing gloves or dry-soaping your hands.

The skylight Bruce made for the Smiths turned out to be less expensive than any other bids they received for a Plexiglas or glass and frame system. It performs well, and adds a dramatic element to the interior design of their house. "Light transmitted through the skylight during the day is broken into the shades of the rainbow," Bruce says, "while nighttime lighting from the house interior shows up the form as strong and opaque." It is no wonder the Smiths are well pleased with it.

Opposite, left: Both sides of the mold for the polyester resin skylight were used to create two layers of glazing.
Opposite, right: A close-up from the house interior shows the copper lattice-work used to establish the shape of the skylight bubble now at work as a design element in the finished window.
Above: The finished skylight sits horizontally on top of a stairwell tower, but does not leak because adequate precautions were taken to seal it from the weather and encourage rain runoff.

155

Megan Timothy

STAINED GLASS FOR ALL WINDOW STYLES

Megan Timothy began making stained glass windows for a very sensible reason: She wanted a small stained glass skylight for her house and was outraged at the price she was asked to pay. She decided to try making one herself. Now, some 400 stained glass projects later, she is still convinced she did the right thing—and urges individuals who come to her wanting a stained glass window or wall hanging to try it themselves, with a little bit of help from her to get them started.

"Where people get bogged down," she says, "is in design. A lot of them could design very well, but they're just not confident that they can. This is what I try to give people when I teach—confidence."

Looking up at Megan's patio ceiling, which acts as a room-sized skylight in stained glass, it is clear that Megan is confident of her own design sense—not only in aesthetic terms, but also in terms of

156

structure and function. Even on one of California's rare rainy days, the patio room is alive with warm light, and as dry as any other room in the house.

There are a number of factors that contribute to the success of the design—not the least of which is the forethought Megan gave the project. "When designing a custom window, I always ask myself, what is the purpose of the piece? Is it to block a view? Frame a view? Enhance a view? Is it to be the focal point of the area? Or is it to be part of its overall surroundings?"

When those questions have been answered, the light question has to be studied. "It is appropriate to use the heaviest types of glass in west-facing windows and more delicate glass in north-facing windows to take advantage of differing amounts of available natural light," Megan says.

When designing a skylight, Megan is especially concerned with color. She believes that yellows, browns, oranges and reds give a warm, glowing light to a room—and enhance the look of the occupants. It is for these reasons that yellows and oranges dominate the palette she used for her patio ceiling.

Although Megan is convinced that intricate stained glass designs are less likely to shatter in earthquake-prone California than designs which use large pieces of glass, she prefers the latter. She outlines her areas of color with just a few simple lines of lead or copper foil and either keeps the overall size of the window small or reinforces it to distribute stresses evenly.

Megan took both of these precautions in designing and installing her patio ceiling—she limited the size of each stained glass window section, and she added plenty of structural reinforcement. Using yellow, orange and white stained glass, she created an abstract design wide enough to span the distance between existing I-beams. She bolstered the middle of each 2 by 3-foot window section by soldering a galvanized steel bar across the leading.

Megan repeated her design over and over again until she had enough stained glass window sections to fit between the I-beams like continuous sheets. Then she climbed a ladder to bead the inside edge of facing I-beams and crossbeam with silicone caulk, and fitted the first window section in place at the roof "peak." One by one, she caulked the I-beams and butted window sections up against each other until all were in place.

To protect the glass from rain, occasional though it may be, Megan double-glazed her patio ceiling. She hoisted big sheets of Plexiglas onto the top of the I-beams and applied them like shingles, caulking along the beams and overlapping sheets. In this fashion, she "roofed over" her stained glass. In the seven years the ceiling has been in place, it has never leaked. The only maintenance required is an annual inspection of the caulking and occasional window-washing to invite in the sunshine.

Of course, someone experimenting with stained glass for the first time will want to try much smaller projects before tackling anything as large as an entire ceiling. However, when you feel confident enough to try a window, don't be afraid to attempt a skylight. Megan feels that a manufactured skylight unit can serve as a great beginning for a stained glass skylight. "Since there is no way to waterproof a near-horizontal

piece of stained glass, you can solve the problem by making your stained glass window section to fit under a prefabricated fiberglass dome," she advises. "These are now easy to purchase in a wide variety of shapes and sizes."

Since most of Megan's clients live in California, she does not often double-glaze for insulation's sake. She does, however, caulk thoroughly and use drip caps and kerfs to help prevent leaks. When she is asked to double-glaze a fixed window for a rectangular or square opening, this is the procedure she follows:

Megan determines the length and width of her stained glass window pattern simply by taking very careful measurements of the rough opening—checking at two or three points the distance from jamb to jamb and head to sill. She makes a template if her measurements show the window opening to be out of true, or if the opening is round or oval or another unusual shape.

Then she assembles her stained glass window, using H-channel lead came around the perimeter rather than the U-channel traditionally recommended. She relies on the accuracy of her measurements and the extra thickness of the H-channel came to make her stained glass window fit. If the window opening is a little out of true, she files the came until the window fits with a fraction of an inch clearance.

If possible, Megan uses the original stops to hold the glass in place. If not, she cuts new inside and outside stops from quarter-round molding. To fit between the layers of double glazing, she cuts rectangular stops from 1-inch stock. She miters the ends, and finishes the

stops to match the interior and exterior casing.

When the finish on the outside stops is dry, she tacks the stops in place lightly with 10d finishing nails, and tests the fit of the exterior glazing. When the fit meets with her approval, she sinks the nails and caulks all around the stop with silicone caulk. (There are many brands of silicone caulk that will do. "I try to find clear silicone that dries kind of rubbery to allow for expansion and contraction. If you caulk with something that dries hard, the caulk may crack with the expansion and contraction of the glass and you'll lose the airtightness you desire.")

Once the silicone caulk has dried, Megan installs the exterior glazing. She uses insulated glass if the homeowners' primary concern is energy-efficiency, ¼-inch tempered plate glass or safety glass if they are more concerned with protection against possible vandalism.

She nails the rectangular stops in place to establish and maintain the insulating air space

Megan Timothy uses stained glass not only in fixed windows, but also casements and double-hung sash, ceilings and shower doors.

With a little ingenuity, Megan Timothy has found ways to double-glaze windows with multiple lights.

between exterior and interior (in this case, stained glass) glazing. Then she cleans the inside of the exterior glazing thoroughly. She considers this her last chance to take care of excess caulk, adhesive backings from manufacturers' labels, dirt streaks and fingerprints before the stained glass goes in place, so she is careful to be thorough.

Next she checks the fit of her stained glass window. She files down the lead came where necessary. She caulks the inside face of the stops, then presses the stained glass window in place against them. Finally, she nails the interior stops in place. Unless the installation is in a game room, she feels the stained glass window will be well protected from possible damage by this double-glazing arrangement. In addition, it makes the window more energy-efficient without sacrificing any of the luminous beauty of the stained glass. This method of double glazing works well in any fixed window—whether it is rectangular or octagonal, round or oval—provided that the stops are miter-cut to length or curved accordingly.

Where temperature differentials make condensation between glazing layers a problem, the moisture can be vented through a tiny hole drilled from the underside of the interior trim to the air space between the layers of glazing.

Single-glazed stained glass windows, both fixed or operable, are even simpler to install. Megan has "picture windows" in both her dining room and living room that consist of a fixed window with a casement window on either side. To frame pretty views, she made her stained glass windows of clear glass with a

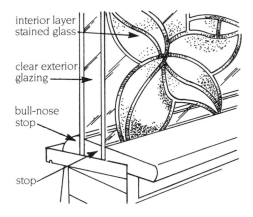

Double-glazing a stained glass window.

interior layer stained glass

clear exterior glazing

bull-nose stop

stop

The most difficult thing about making and installing this round window was finding a company that manufactured window stays. Construction of a simple wooden window stay is explained on page 125.

perimeter pattern of colored glass and leading.

Megan sealed the center window, which is bordered in H-section came, against the original stops with silicone caulk. Then she cleaned it and tacked inside quarter-round stops in place against it. She installed her stained glass in the casement windows in the same way.

Megan is not at all apologetic about the simplicity of her installation. "The weather is always nice," says this adoptive Californian, "except for about two days a year. I don't have screens and don't have storm window inserts. They're just not necessary. I just leave the casements open to the air most of the time.

"For each part of the country, though, you have to adapt to the environment. If I lived in Louisiana, I would not be without a screen in the window because they've got mosquitoes as big as condors!" (In case you *do* live in Louisiana—or Alaska or New Jersey—you can install a screen or storm window insert on a stained glass casement window just as you would on a standard casement window.)

Megan has also designed and installed stained glass in double-hung windows. In working on the hummingbird window, shown at left, she had to take into account not only the thickness of the parting rail, but also the thickness of the muntins. Because the individual lights are so small, she used copper foil to join the pattern pieces and left off the border came.

To install each light, she applied glazing compound against the outside stops, pressed the light in place, and pressed in glazing points. Then she fitted inside stops against the light. After waiting for the compound to dry, she trimmed away the excess with a razor and washed the glass. Megan cautions that this easy installation is possible only because individual lights are so small — 4 inches square. Above this size, the structural support of border came is essential.

An alternative to side-hinged casement and double-hung windows that also allows ventilation is the center-pivoting window. Megan finds this installation method works particularly well for round windows, although it can be used with more traditional shapes.

To make a center-pivoting window, she installs her stained glass in its own wooden frame. She attaches the pivot hardware on opposite sides of the sash along the horizontal or vertical centerline, depending on how she wants the window to swing. She uses a simple hook-and-eye latch to hold the window closed or shops for a more sophisticated locking mechanism.

To hold a center-pivoting window open requires a little more ingenuity — or a willingness to do long-distance mail-order shopping. Megan, who grew up in South Africa and has spent a lot of time in the British Isles, orders brass window stays from a firm in London. The stay has a metal bar that attaches to the bottom of a center-pivoting window. To hold the window open, the bar fits over an upright peg attached to the windowsill. Such devices — still quite common in older houses abroad — predate the crank-handle mechanisms to which most Americans are accustomed.

Megan's innovative use of stained glass doesn't stop with windows. She also does countertops, backsplashes and shower doors. "I'm told I was the first stained glass artist to introduce stained glass shower doors," she says proudly. "I have had two of these installations in my own home for over six years, in use every day and not even one crack in them."

When making a stained glass panel for a shower door, Megan is careful to use the sturdier glasses — glue chip, cathedral and opalescent. To add structural support she solders two or three brass or steel brace bars across the inside of the finished stained glass panel.

Otherwise, she treats the panel like ordinary replacement glass. She sizes it to fit a standard aluminum shower door frame, and notches the frame to receive the extra thickness of the brace bars. Then she caulks the inside of the aluminum channel and fits the panel in place.

The greatest advantage of using a standard aluminum shower door frame is that such frames come with their own special sills — a rubber gasket or a sliding track — either of which will prevent water from leaking out from under the shower door.

Anyone tempted to make a stained glass panel for a shower door should know that in most parts of the country they are illegal, since no building codes exist to set standards for them. As a result, Megan is careful where she installs them. "I could not put a stained glass shower door in a rental house, because if there were an accident, I could have a big problem. If I were to sell my own house, there would be no mention of a stained glass shower door in the sales contract. I would take the stained glass shower door off and leave it in the garage. I would install a regular shower door. What the people who bought the house did with the stained glass shower door after I left would be none of my business."

Megan believes that her installations are just as safe as standard shower doors. "The stained glass doesn't shatter when it is braced unless you really hit it. I think if you fell against it, you would merely crack it. To make it shatter you would have to hit it with a hard object like a hammer, and give it a real whack."

Megan is quite comfortable keeping such practical considerations as building codes, moisture leakage and energy-efficiency in mind when she works in stained glass—and that is as it should be. Unlike a painting or a piece of sculpture, which exists only to be contemplated and admired, a stained glass window or door is a work of art with functions to perform. Megan Timothy's designs succeed not only because they add beauty to their surroundings, but also because they are soundly made and installed with care to insure a long and useful life.

Stained glass adds beauty in every room of Megan Timothy's house—even the bathroom and dressing room.

163

FACETED GLASS WINDOWS

Playing with the products of space-age technology, using rigorously engineered woodworking and glassworking tools and power equipment, Bruce Sherman explores a new craft form: three-dimensional, faceted glass sculpture.

He is quick to point out that he did not invent the process he now uses to create beautiful, crystallike windows, skylights, lighting fixtures and mirror sculptures. He simply uses it on a grand scale, and with the vision of an artist.

Bruce learned how to create faceted forms with glass and silicone from Kim Hicks and Lou Galetti, free-lance designers he met while working at Western Woodcraft, a large woodworking shop. They taught Bruce the basics and encouraged him to go on his own and try large-scale applications.

With a degree in architectural design and drafting from the College of San Mateo and experience as an architectural

164

modelmaker, Bruce felt confident of his design skills, but not of his entrepreneurial flair. "When I started out in January, 1972," he says, "I was just a naive guy trying to find my way." He managed to keep himself fed with modelmaking, and within a few months, found time and patrons enough to allow him to experiment with faceted glass constructions. With word of mouth as his only advertising, he has been busy with commissions ever since.

Bruce is usually at work by 9:00 or 9:30 — but not behind a desk or on an assembly line. He does most of his design work at a coffeehouse. "The owner tolerates me well," he says. "He's a wonderful guy. He has the most wonderful employees, too — cheerful."

So, sipping espresso, he opens his briefcase and unpacks his drafting instruments and graph paper. He begins his design work by doing freehand drawings. He takes into account the client's needs and taste, his own impressions after a visit to the site, and any rough frame measurements he was able to get. He sketches both front and side views of the faceted windows until he is pleased with the pattern of line and form. Finally, often after days of refinement, he makes precise scale drawings of front, side and corner views. If it seems appropriate to use colored glass, he indicates it in the drawing.

He returns to the client for approval of the design, and to discuss his fee. "Generally, I request 50 percent of my payment at the beginning of construction and the remainder upon completion," he explains. "Most are amenable to this, bless their hearts."

Back home, he works up full-scale drawings showing all the wooden frame members

and joints. He is very careful to be precise. "Since the glasswork is only as strong as the frame, I spend considerable effort to build as strong a frame as possible, allowing for movements through the years as the wood expands and contracts with changes in the weather." His considerable effort includes carefully selecting close-grained redwood lumber, notching joints so that they fit together "like a Chinese puzzle," reinforcing the joints with screws (hidden by wooden plugs), and using yellow (aliphatic resin) glue and silicone caulk.

Since he has only one room in which to do both woodwork and glasswork, he completes his woodworking first, all the way from milling the lumber to staining or painting the frame. He interlocks all frame members and rabbets the inside edges to hold the glass. He dry-assembles the frame in his shop to take accurate measurements for his glasswork, and then works up full-scale front, side and auxiliary views of each of the glass facets.

To begin the development of each facet he needs the measurement of at least one edge. Then, using the theorems of descriptive geometry, he can arrive at the other measurements. Often he uses the length of a facet along the frame to begin the process. Once he has all the dimensions of one facet, he can proceed to figure out the dimensions of the facet adjacent to it. Just as an answer to a crossword puzzle that

Bruce Sherman's faceted glass windows can take unexpected forms because he assembles them in glass and wood modules that interlock for strong structural support.

reads across gives one of the letters for an answer that reads down, when Bruce determines the length and location of all the edges for one facet, he has the information he needs to develop another. Although the process is logical, it is not always easy.

"I make my drawing with a needle-sharp drafting pencil and use stick pins to mark line intersections. I have it taped to a parallel-bar-type drafting table as I wend my way through the descriptive geometric process of developing the true full-size shapes of all the facets. I view the drawing under a magnifying lens and strong light. From one view, say, the front view, I might have the true *length* of the edges which sit in the wooden frame, but since the glass comes away from the wood at an angle, I need to find the *width* of the facet from an end-section view which shows the facet merely as a single edge line. It goes on and on like this, and gets worse in oblique facets, which have no easily found true-length edges in any of the views I've drawn so far."

After Bruce has arrived at the dimensions of all the facets this way, he makes a template drawing of each one, and tapes the drawing on a simple horizontal working surface set up with a special glass-cutting jig. One of the lines on the template drawing (indicating a reference edge the facet holds in common with another facet) goes under a Plexiglas stop.

He places glass over the drawing, and braces it against the stop. He uses a straightedge to guide his diamond-tipped glass cutter as it scores the glass. Then he breaks the glass along the score line. He sands the newly cut edge of the glass (on a wet belt sander with a silicon carbide belt) ever so slightly to make it safer to handle.

After all the glass is cut and ground, he covers what will be the outside surface with contact paper or masking tape. Then he begins to tape together the facets of his three-dimensional windows. As much as possible, he works in modules. For example, the 10-foot-high stairwell window was built in three modules, joined together on the job. Each bay of the six-bayed kitchen window was built with two modules — one of all the vertical glass, another of the near-horizontal glass. The band of faceted glass connecting them was glued in place on the job.

Making a module in his shop requires that Bruce build a supporting structure, "like scaffolding for a freeway overpass under construction." He uses cardboard or wood to shape the contour of each window section. Then he joins the facets of the glass, inside edge to inside edge, with masking tape, and drapes them over his supporting structure. He beads the silicone glue into the joint, between the exterior edges.

The glue Bruce uses is neither difficult to get nor expensive. You can buy Dow Corning RTV-732 silicone rubber glue in most hardware stores. Made for use on space satellites, it is moisture and temperature stable and somewhat flexible once it has vulcanized, a process that begins as soon as it comes out of the tube. At room temperature, it reacts with moisture in the air to form translucent silicone rubber. A byproduct of the chemical reaction is the formation of acetic acid, so if you decide to try making a faceted glass window, be prepared for the smell of vinegar when you apply the glue.

"Right after the glue is squeezed in between the glass edges," Bruce cautions, "it must be smeared smooth with a finger or palette knife." After the glue has set overnight, Bruce removes the masking tape from the underside of the joints and cleans away any excess glue with a razor blade. He polishes the glass along the glue lines by sprinkling on pumice powder and rubbing with a soft, moist cloth.

Depending upon the complexity of the installation, Bruce may glue up each frame section that holds a glass module in his shop and join it to similar sections on the job, or he may glue up the entire frame and set the glass modules into it on site. He carries the works in the hatch and on the roof rack of his small Datsun, padding the glass modules with foam rubber before strapping them down.

At the job site, Bruce fits the sash into the rough opening left for him by the carpenters. In the case of the pentagonal kitchen window, he glued the glass modules into the frame parts on the kitchen floor. He applied silicone rubber glue to the rabbets on the frames and then set each module in place so that the edges of the glass rested on this bed of silicone rubber. He let the glue set overnight and got carpenters to help him lift the 150-pound sash into the opening. He secured the sash to the wall studs with screws, added trim, and caulked with more silicone rubber around the perimeter to help prevent leaks.

Bruce's precise measurements and careful joinery have made leaks a rare problem. However, if an air bubble in the silicone rubber glue escapes his notice and is squeezed into a joint,

the window will develop a tiny leak. To fix it, Bruce simply applies another dab of the glue.

Perhaps not surprisingly, this inventive craftsman has not been content to build windows with only convex forms. He has successfully experimented with convex and concave combinations in tall stairwell windows. There are two "trade secrets" involved: Whether part of concave or convex forms, the facets must meet at an angle between 90 and 180 degrees (optimum is 120 degrees) for greatest structural stability. And, since the panes of glass in a concave form come nearest to touching on the exterior surface, the glue must be applied from the interior, to join the panes of glass where the gap between them is widest.

After nine years of observing his faceted-glass windows in both the California sunshine and the area's severe winter rains, Bruce is quite confident of their structural stability. He thinks of his silicone rubber joints as long hinges, flexible enough to expand and contract with moisture and temperature changes in the air and sturdy enough to hold up against high gusting winds.

Bruce uses smooth or textured clear glass almost exclusively, and adds stained glass sparingly, usually only at the customer's request. He feels that the natural beauty of the crystal forms needs nothing to augment it but a view. Anyone who has peered through a kaleidoscope and wished to capture its luminous patterns of mirrored colors can understand his viewpoint; those who have seen his windows, dancing with reflections of San Francisco's blue sky and flower-filled gardens, would agree.

The modular construction Bruce Sherman uses to build his faceted glass windows gives him flexibility to work even on a very large scale, as this stairwell window demonstrates.

Lynn Kraft

SAND-BLASTED GLASS WINDOWS

The old Victorian house, set scarcely 25 feet off a busy highway, was dirty from years of road grime. It had been vandalized repeatedly, and pigeons fluttered in and out of the open third-story window. Most people would have passed it up without a second thought, but when Lynn Kraft saw the "For Sale" sign in the front yard, he jumped at the chance to buy the property.

When Lynn bought the house in 1973, he was fresh out of college with a degree in art education and living on the low wages of his first teaching job. He risked his savings as a down payment, took on a mortgage, and promised himself he would restore and remodel the house only as he could afford it—one project at a time. For two years, he lived there on a makeshift basis, working on the house at odd hours after school and on the weekends. In 1975, when he got a working heating system installed, he and his brother Gordon moved in

170

for good. Since then, the brothers have worked together renovating the house, and it is their love of craftsmanship that has transformed it into the beauty it is today.

One of Lynn's early projects was to replace the house's many broken and boarded-up windows. "I wanted the windows to be special," he says, "so I decided I would try etching the glass myself. I didn't have any experience in working with glass in any of my classes at art school, and I didn't have anyone to teach me, but I've always been willing to experiment and learn things for myself. I did a little research, and went out and got the equipment I needed."

Lynn thought about the design for the windows of the double entry doors for some time. "I always wanted a house with a name," he says. "I thought of the mill race, and the thorns and briers that covered the hill when I first bought the house, and I tried to find one word that summed up all those images." His final design featured the word "Thornbrook" on one window pane, and a delicate thorned branch with berries on the other.

Lynn felt the design was very pretty—but too delicate. It just didn't have the impact he wanted. Besides, he thought the process of hand-etching with scribing pen and motor tool too laborious and slow to use for other windows in the house.

He began to explore other methods of achieving an etched glass effect. A local glass supplier gave him a small container of hydrofluoric acid, which is commonly used by glass manufacturers to create frosted glass. Because even the acid's fumes are very dangerous, he had to take his project outside. The acid produced the matte surface he wanted, but it also leaked under the wax resist he used, blurring the edges of the design. He was not happy with either the method or the results, so he decided to keep looking for a better way.

The discovery of the etching method he now uses was inadvertent. Over one summer, he decided to sandblast the front of the house to remove the dust and grime that had accumulated on the brick from all the traffic on the highway. "When I rented the sandblaster and air compressor," he says, "the salesperson warned me, 'don't let the sand touch glass.' Naturally, the first thing I did when I got the equipment home was to find a scrap piece of glass and begin to experiment.

"I loved the evenness and sparkle the sandblasting created. I immediately decided I wanted etched glass windows all over the house, like a Victorian ice cream parlor." True to his principle of one project at a time, however, he decided to begin with the half-round window under the central tower, which admits light into a stairwell. Since the architecture is somewhat formal, Lynn sketched a symmetrical design for the window.

The half-round window, which had been set in a removable frame, had been broken and the gap covered with cardboard. To make a window pane for the opening, he had the local glass supplier cut ¼-inch plate glass to fit. Since

sandblasting is a process of controlled erosion, Lynn needed to cover those areas of the glass he wanted to remain clear with a material that the sand grains would not easily penetrate. He used layers of masking tape as a resist. To transfer his design to the masking-taped surface, he inserted carbon paper under the paper on which he had drawn the design and retraced the design. He then cut away sections of the masking tape with an X-Acto knife, exposing the areas of glass he wanted to sandblast.

The actual sandblasting was a relatively simple process. To protect himself from being cut by tiny shards of flying glass and grains of sand, Lynn wore goggles, mask and gloves, and worked outside. Using a very fine grade (00) of sand, he turned the air compressor to 60 pounds of pressure and blasted the prepared glass. The effect on the half-round was as elegant and dramatic as he could have wished.

More experiments followed. By the time Lynn did the second-story windows he was using a double sheet of contact paper as his resist. Since the original glass was in place, but broken, it had to be replaced. To remove it, Lynn chipped away the putty on the exterior, pulled out the glazing points with pliers, and lifted out the old glass. After he sandblasted his design onto replacement glass, he reglazed the windows. So pleased was he with the shimmer and sparkle of his sandblasted windows that it was not long before he had replaced the glazing in the entry doors that he had so painstakingly hand-etched.

Introducing the students in his high school art classes to his techniques has brought refinements and innovations. The process Lynn sug-

Lynn Kraft sandblasted an ornate design on his windows to give a feeling of Victorian elegance to the century-old house he is restoring and remodeling, one task at a time.

gests his students follow begins with cleaning the sheet of glass thoroughly and allowing it to dry. Next, two sheets of contact paper are applied to its surface, and any air bubbles caught under the contact paper are punctured with a pin and smoothed out. The design is transferred to the contact paper using carbon paper, 20-pound drawing paper, and a ballpoint pen, then cut out with a sharp knife. The contact paper on the areas of the glass to be frosted is peeled back, and the exposed areas are sandblasted.

To make it possible to safely sandblast glass inside the art room at the high school without wearing goggles or mask, Lynn built a glove box. The box, which measures 40 by 30 by 48 inches, rests on a counter 30 inches high. The top and the front of the box act as hinged doors so that even a large sheet of glass can be lifted in or out easily. Against the back wall of the box, a shelf slopes into a trough for the sand grains. A fabric sleeve funnels the sand in the

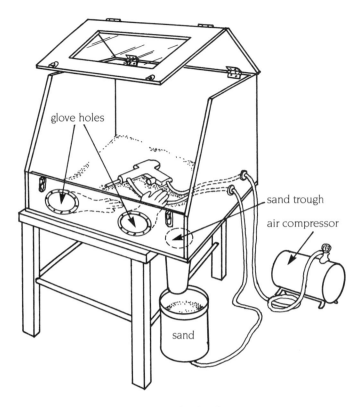

glove holes

sand trough

air compressor

sand

The glove box Lynn Kraft made for his high school students to use when sandblasting glass.

trough back into its container on the floor.

The hoses for the sandblasting gun come in through holes in the side of the box. The front of the box has a window in it that allows a view of the glass being blasted. Holes in the wooden frame under this window permit manipulation of the glass and the sandblasting gun by the operator, who wears muslin-sleeved gloves, which are stapled to the frame. Two utility lights in the top of the box make it easier to see if the glass has been thoroughly blasted.

Sandblasting glass is a simple matter of setting the air compressor at 45-60 pounds per square inch, pointing the sandblasting gun at the prepared surface, and pulling the trigger. Lynn and his students have discovered that it is important to be thorough, since spots that haven't been sufficiently blasted will reflect light differently and show up as flaws. He advises his students to blast the exposed glass in slow, overlapping circles to create an even, frosted effect. If bad spots show, Lynn recommends marking them with a pencil and reblasting. When the desired effect has been achieved, the glass is lifted out of the box, all the contact paper is removed (warm water helps), the glass is cleaned and dried, and its edges are taped or framed.

Lynn's students have willingly experimented with resists other than contact paper. They have found that layers of newsprint, rubber bands, beads of Elmer's Glue and yarn can be used successfully to vary the texture of the surface and the form of the design.

They have also experimented with placement of the patterns. Sandblasting one pattern, and then moving it or using a second pattern to sandblast the same area of glass varies the depth of the sandblasting, so that in some areas the design is "whiter" and less light-transmitting.

Although the sandblasting eventually wears down the pattern, designs can be used again. Lynn used the design for the half-round window a second time when a strong wind blew out the original sandblasted window, which had been secured in the frame by only one nail. (He has since learned to be much more careful about proper installation.)

Now he is thinking of double-glazing some windows before the coming winter. Each of the double entry doors is thick enough that he can remove the inside window trim, bore a tiny hole in the lock rail and fill it with silica gel to absorb moisture from condensation between the glazing layers. He will install the clear exterior glazing, add a window stop, and then install his etched glass window pane. Glazing points and inside window trim will hold it in place. Finally, he will use a silicone caulk to weather seal both layers of glazing in each of the double entry doors. This won't detract from the beauty of the windows, and it will add some warmth to the entry hall, where so many people have said their hellos and goodbyes to Lynn and Gordon Kraft.

Those who once find their way to "Thornbrook" do not soon forget it. As the windows in the double entry doors shimmer with the light of the late afternoon sun, the lacy designs etched in them cast intricate shadows on the wall—and in memory.

"I always wanted a house with a name," says Lynn Kraft—so he made one up and sandblasted it across the windows of his entry doors.

Joe Devlin

SHEET METAL WINDOW GRATES

By training and occupation, Joe Devlin is a scientist, comfortable in computer rooms and testing laboratories. By choice, he is a do-it-yourselfer. A few years ago, he bought a crumbling three-story brick house in the Brewerytown section of Philadelphia. On the evenings and weekends, he worked to bring its electrical and wiring systems into the twentieth century and to make necessary structural repairs. Two weeks after he took the boards off the first-floor windows, repaired the window casings, and replaced the glass, the house was broken into and robbed. The boards went back up, not to come down again until Joe came up with a way to make his first-floor windows secure.

Joe investigated manufactured window grates, but was not impressed with the available selection. Permanent vertical bars and expandable safety grates seemed institutional and grim. Factory-cast ironwork was

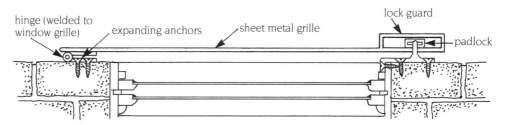

Joe Devlin's window grille design, top view cross section.

hinge (welded to window grille)　　expanding anchors　　sheet metal grille　　lock guard　　padlock

more decorative, but still forbidding looking when installed. Since the costs of any of these grates was prohibitively high, Joe didn't even attempt to commission custom work. He decided, as he so often does, to do it himself.

He had a little bit of experience in electrical welding, gained as he worked his way through college, which gave him the confidence to try to make his own window grates. He had enjoyed the work and had been tempted ever since to buy his own oxyacetylene welding rig. Now, with this project in mind, he plunked the money down. "An oxyacetylene welder," he explains, "is basically two tanks (one for oxygen, one for acetylene), two hoses that connect the tanks to a regulator, and another double hose that connects the regulators to the torch." This kind of welding rig is available through the Sears catalog.

Joe salvaged the steel sheet metal for one of his grates from a roadside ditch and for the others from dumps at construction sites. He doesn't know what purpose the metal he used for the oak tree grate was originally intended to serve, but it was heavy—a couple of hundred pounds—and awkward to haul out of the ditch and into his car. The metal he used for the two ibex grates was part of the housing for a large electrical transformer that had been dismantled. It was not as thick as he would have liked, so he reinforced it with sections from the metal framework of a bed's box spring. Joe's father has been involved in the salvage business for years, so Joe has learned to keep his eyes open for abandoned

furniture that might have a second life. The box spring was found on Joe's weekly Friday afternoon trash survey.

Joe wanted the grate designs to be decorative and in keeping with the decor of the rooms from which they would be seen. He was also concerned that they be substantial enough to discourage even determined burglars. He decided to make the grates for the 6-foot, double-hung windows high enough to protect the lower sash and the parting rail, but not so high that they kept daylight from the house. He settled on large silhouettes with delicate edgework.

He found the oak tree design used in his kitchen window grate on the wall of a department store restaurant in downtown Philadelphia. He paid a second visit to the restaurant with his camera to record his inspiration. With photo in hand, he drew the design on tracing paper, adapting it to his special needs.

The design for the ibex, a long-horned goat native to the mountains of Turkey, came from a library book of animal silhouettes. Joe traced the silhouette and modified it to come up with a design that fit his requirements.

A photo enlarger was used to blow up tracing paper sketches, but anyone without this equipment can enlarge design sketches using a proportional grid system. Joe cautions that "the final, full-size pattern must be large enough to overlap the sides of the window opening, allowing room to attach the hinges on one side and to padlock the grate to the wall on the other side."

177

Joe Devlin made his protective window grilles attractive by cutting delicate silhouettes in sheet metal with an oxyacetylene torch.

Joe cut full-size cardboard templates, then outlined them on the steel with soapstone. On black steel, soapstone leaves a thin but visible line.

He set up his welding rig in an unused room on the second floor. With a steel sheet underneath him to protect the flooring, he practiced on metal scraps until he felt he could work fairly accurately. Then he cut out the silhouettes. The ⅛-inch steel of the oak tree grate didn't need structural reinforcement, but the 3/32-inch steel of the ibex grates did. So Joe cut metal from the salvaged bed box spring, and welded a rigid horizontal bar across the width of the ibex and a reinforcing border around the silhouette outline. Then he added "lock guards" to both grates.

The locking mechanism Joe invented to protect the lock is simple but quite effective. A metal latch is welded to the grate's edge at a 90-degree angle to its decorative face and is drilled with a hole large enough for a padlock. An L-shaped latch keeper is fastened to the wall parallel to the decorative face of the grate (which has an opening wide enough to receive the latch). Protecting the latch, latch keeper and padlock is a lock box, which is incorporated into the decorative face of the grate. The arrangement allows the grate to be locked and unlocked from the inside by opening the window. Joe keeps the key on a hook near the window to make a quick exit possible in case of fire.

Mounting the grate on the house requires a little knowledge of its construction. Joe suggests using expanding anchors to secure the hinges to brick, concrete or cinder block. Lag bolts, long

screws or through-the-wall nut and bolt combinations will work well on frame construction. Of course, it is important that the hinges be placed so that the hinge pin and mounting screws are behind the closed grate, out of reach of prying hands.

Joe is pleased with the product of his labors: "I think the time and effort invested in building my window grates has been well spent. I believe the grates blend well with the atmosphere of the rooms they are in. I know I feel more secure now that they stand between my home and those who would intrude."

Blacksmith Gregg Leavitt forged this ornamental ironwork grille, left, *to provide some security for the homeowners. Sculptor Jack Boyd handcrafted this grille*, right, *for the entry door of his studio and shop, making the designs most intricate over the area of glass near the door handle to prevent unwanted intruders from simply breaking the glass and letting themselves in.*

Tom Bender

ROLL-AWAY WINDOW WALL

Tom Bender and Lane deMoll are the kind of people who live by their convictions. Together, they designed an energy-efficient house that expresses their feelings about shelter and open spaces. Tom, who trained as an architect at the University of Pennsylvania, and Lane, a writer for *RAIN: The Journal of Appropriate Technology*, wanted their house to be no more of a barrier than necessary between them and the world around them.

Travel to Japan, India and Persia convinced them that their house could connect them more closely to nature instead of separating them from it. They learned to love the Japanese garden pavilions whose paper screen walls can be moved aside to invite spring breezes into the house. They grew to understand why the Japanese value that same intimacy with nature in winter, when insulated clothes, rather than thick walls, provide warmth

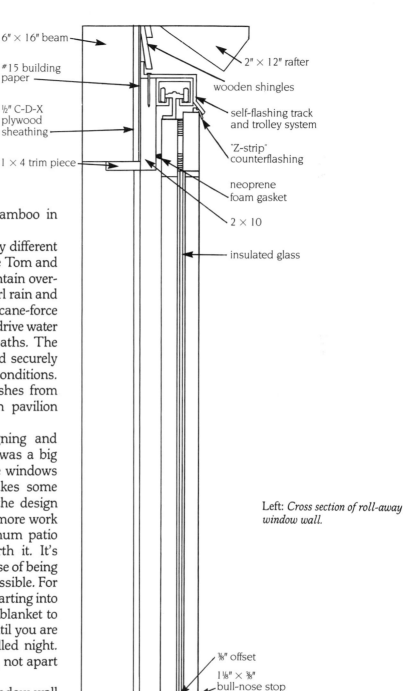

6" × 16" beam

#15 building paper

½" C-D-X plywood sheathing

1 × 4 trim piece

2" × 12" rafter

wooden shingles

self-flashing track and trolley system

"Z-strip" counterflashing

neoprene foam gasket

2 × 10

insulated glass

⅜" offset

1⅛" × ⅜" bull-nose stop

½" × 1½" filler strip

2 × 6 bottom rails

neoprene foam gasket

flooring

Left: *Cross section of roll-away window wall.*

as the snow falls softly onto the bamboo in the garden.

Of course, Japan's climate is very different from that of Nehalem, Oregon, where Tom and Lane built high on the side of a mountain overlooking the Pacific. Winter storms hurl rain and salt spray at the house with hurricane-force winds. Gusts up to 100 miles an hour drive water through even the most unexpected paths. The window wall Tom and Lane designed securely seals shut under these severe winter conditions. But on summer days, the wall vanishes from sight. The house becomes an open pavilion high above the waves.

Tom acknowledges that designing and building the roll-away window wall was a big project, but he says, "Like any of the windows we've built, the first-time design takes some work. But once done, you can use the design over and over. This design took a lot more work than slapping in a couple of aluminum patio doors, but the result was well worth it. It's difficult to describe the wonderful sense of being outside *inside* a room that it makes possible. For that, you have to see hummingbirds darting into the room to pluck a bit of wool off a blanket to line their nest, or sit in the twilight until you are surrounded by the hush of a star-filled night. You are very simply a part of all that, not apart from it."

Tom and Lane designed the window wall with sun angle, wind direction and rainfall in mind. They placed it on the south side of the house for the greatest year-round warmth and protection from winds and rain, and sloped the roof overhang above it to shade the glass from the summer sun. Smaller operable windows

installed across from the window wall permit cross-ventilation. "Whatever your orientation and climate," Tom cautions, "it is essential to keep air movement and temperature within reasonable limits inside the room or you'll find that you will never open your window wall."

Tom and Lane's design consists of two 9-foot-wide window wall sections suspended from an overhead track *outside* of the house. This allows Tom and Lane and their son Skye to sit on the edge of the living room floor without having hot and uncomfortable metal tracks under their thighs.

Tom built the window wall sections of construction-grade lumber and standard 34 by 76-inch sheets of tempered patio door glass. He needed the help of three friends to lift the 6-inch by 16-inch by 18-foot fir beam that spans the opening into place. He was able to do most of the remaining construction himself.

"When we started building, we pulled enough straight and clear 2 × 4s and 2 × 6s for the window wall out of the pile of framing lumber and stacked it aside to dry. If you do that," Tom cautions, "be sure to set aside enough extra to allow for the inevitable ones that develop some twist or bow as they dry.

"When we were ready to construct the window units we sorted and coded the wood for each particular part. Straightest 2 × 4s were allocated for the end verticals, while others were used in between where the glass and joints would help keep them true. Similarly, the 2 × 6s were placed so that any bow would be on the inside where it could help the latches pull the whole 9-foot length of each window wall section in snug, and any camber in the boards would counter eventual sagging."

Tom rabbeted the 2 × 4s used for sash stiles and mullions and offset the pairs of 2 × 6s used as sash rails to receive the glass and stops. Then he notched out the joints and assembled the sash frame with screws and glue.

He bolted the self-flashing track to the beam and fastened trolleys to the sash rails, notching the rails wherever necessary to fit the trolley brackets. Then, with the help of a friend, he hung the unglazed window walls and made the necessary adjustments to make them hang true. A windowsill that projected into the path of one unit was trimmed, and a ½-inch-thick metal bumper was attached to the exterior walls to prevent the bottom of the window wall from scraping the cedar shingles.

Glazing was then installed, caulked with silicone sealant, and stopped with bull-nose

Putting in movable insulation requires securing the window wall, and popping in the cloth-covered Thermax boards to cover the glazing area within the window frame.

stops. The finished units received a good soaking with clear Cuprinol wood preservative.

Tom knew that closing the window wall against those winter storms would require a real "battening down of the hatches," so he used a combination of gaskets and latches to snug the window wall sections tight against the house. First he screwed a strip of aluminum "Z" flashing in place above the outer top rails to form a spring tension seal against the track. Then he attached a ½-inch tubular neoprene gasket all around the perimeters of the two window wall sections. (This weather stripping acts very much like a gasket on a refrigerator door to form a tight seal against air and moisture infiltration.) Finally, he added his hardware.

A well-stocked hardware store turned up some Nielson VHC 250–25 latches, made like oversize trunk latches. Installed on the sides and top of the frame, these cinch the window units against the house, compressing and sealing the gasket. A V-shaped bracket, welded by a friend, wedges the bottom edge of the windows against the floor, where latches would have been impractical.

Tom and Lane use movable insulation to boost the wall's R-value during the cold winter months. Using a kitchen knife, Tom cut 1-inch-thick rigid foam insulation (Thermax) ¼ inch

smaller than the dimensions of the glazing opening. He cemented vinyl wall-covering material to the insulation panels. These panels fit snugly between the 2 × 4 mullions, with no need for fasteners. When the temperature drops and storms rage, the movable insulation covering the large expanse of glass in the window wall keeps the house cozy and warm.

Tom and Lane don't miss the open pavilions of Japan now. Their roll-away window wall gives their home the same closeness to nature whenever they want it—but keeps them dry and warm during Oregon winters. "We've gotten so we hate to close the window walls even on a cool day," Tom admits, "but we usually close them at night, as raccoons and other creatures get attracted to the kitchen. If we lived in another climate, where mosquitoes and other insects make outdoor living uncomfortable, we'd have to work out a different kind of solution." The window wall gives them such pleasure in their surroundings that it seems likely they would find a way to make it work, no matter where they lived.

APPENDICES

If you want to make your own doors and windows, you will find valuable resources in these appendices. Using them, you should be able to make almost any door or window that you can imagine.

The lists of suppliers include manufacturers and mail-order distributors of tools, native and exotic woods, veneer and inlay, wooden moldings and decorative medallions, paints, stains and finishes, manufactured door hardware, handwrought and handforged door hardware, manufactured window hardware, glassworking materials and tools, glass, and beveling equipment.

Following these you'll find lists of books and magazines which contain how-to information or visual inspiration to help you plan and carry out your door or window design.

Even if you have no intention of undertaking the design and

building of a door or window, you may have learned to admire the craftsmanship and value the uniqueness of the doors and windows photographed for the pages of this book. If you would like to commission a handcrafted door or window, you'll find the craftsmen, custom builders, and custom doormakers and windowmakers listed in these appendices willing and able to make something special for you.

Before deciding who you want to have build a door or window for you, however, you may want to study the photographs in the book and refer to the photo and design credits. In some cases, the name and/or the whereabouts of the doormaker or windowmaker are unknown, but usually you can find the address you want in the list provided.

TOOLS

Contact the tool manufacturers listed to find out where they distribute their products for retail sale in your area. Some may be willing to sell and ship directly to you.

The companies listed as mail-order suppliers do not always manufacture the items they advertise in their catalog, but they usually stock a good selection of both hand and power woodworking tools for sale and shipment anywhere in the country. Often they supply veneers, inlays, moldings, and stains and finishes. Some of these companies (like Brookstone and Garrett Wade) charge a nominal fee for their catalog, but the information they supply on how to use the items they stock makes the catalogs valuable reference works for any woodworker.

Adjustable Clamp Co.
417 N. Ashland Ave.
Chicago, IL 60622

American Machine & Tool Co.
4th Ave & Spring St.
Royersford, PA 19468

Arco Products Co.
110 W. Sheffield Ave.
Graselwood, NJ 07663

Arrow Fastener Co.
271 Mayhill St.
Saddlebrook, NJ 07663

Black & Decker Mfg. Co.
Towson, MD 21204

Brookstone Co.
126 Vose Farm Rd.
Peterborough, NH 03458
(mail order)

Albert Constantine & Son, Inc.
2050 Eastchester Rd.
Bronx, NY 10461
(mail order)

Craftsman Wood Service Company
2729 S. Mary St.
Chicago, IL 60608

The Cutting Edge
3871 Grand View Blvd.
West Los Angeles, CA 90025
(mail order)

Frog Tool Co., Ltd.
700 W. Jackson Blvd.
Chicago, IL 60606
(mail order)

The Japan Woodworker
1004 Central Ave.
Alameda, CA 94501
(mail order)

Lee Valley Tools, Ltd.
P.O. Box 6295
Station J
Ottawa, ON K2A 1T4
Canada
(mail order)

Skil Corp.
5033 N. Elston Ave.
Chicago, IL 60630

Stanley Tools Ltd.
600 Myrtle St.
New Britain, CT 06050

Universal Clamp Corp.
6905 Cedars Ave.
Van Nuys, CA 91405
(mail order)

Garrett Wade
161 Ave. of the Americas
New York, NY 10013
(mail order)

Woodcraft Supply Corp.
313 Montvale Ave.
Woburn, MA 01801
(mail order)

The Woodworkers' Store
21801 Industrial Blvd.
Rogers, MN 55374
(mail order)

MATERIALS

Native and Exotic Woods

All of the suppliers listed below will ship lumber orders directly to customers in other parts of the country. Some have minimum order requirements, however.

Maurice L. Condon, Inc.
248 Ferres Ave.
White Plains, NY 10603

Croy Marietta
P.O. Box 643
121 Pike St.
Marietta, OH 45750

General Woodcraft
100 Blinman St.
New London, CT 06320

Hallelujah Redwood Products
39500 Comptche Rd.
Mendocino, CA 95460

The House of Teak
Chester B. Stem, Inc.
2708 Grant Line Rd.
New Albany, IN 47150

Leonard Lumber Co.
P.O. Box 2396
Branford, CT 06405

Mountain Lumber Company
P.O. Box 285
1327 Carlton Ave.
Charlottesville, VA 22902

Native American Hardwoods
RD #1
West Valley, NY 14171

The Sawmill
The C. F. Martin Organization
P.O. Box 329
Nazareth, PA 18064

Sterling Hardwoods, Inc.
412 Pine St.
Burlington, VT 05401

TECH Plywood & Hardwood
    Lumber Co.
110 Webb St.
Hamden, CT 06511

Unicorn Universal Woods, Inc.
137 John St.
Toronto, ON M5V 2E4
Canada

Weird Wood
Box 190 FW
Chester, VT 05143

Willard Bros. Woodcutters
300 Basin St.
Trenton, NJ 08619

Wood Shed
1807 Elmwood Ave.
Dept. 9
Buffalo, NY 14207

Wood World
9006 Waukegan Rd.
Morton Grove, IL 60053

Veneer and Inlay

The companies listed below are mail-order suppliers. In addition, see the mail-order woodworkers' supply companies listed under "Tools," especially Albert Constantine & Son, Inc. and The Woodworkers' Store.

Homecraft Veneer
901 West Way
Latrobe, PA 15650

Bob Morgan Woodworking Supplies
1123 Bardtown Rd.
Louisville, KY 40204

Wood Moldings and Decorative Medallions

All of the companies listed below are mail-order suppliers. See also mail-order woodworkers' supply companies listed under "Tools."

Cumberland Woodcraft Co.
RD #5, Box 452
Carlisle, PA 17013

Driwood Mouldings
P.O. Box 1729
Florence, SC 29503

Focal Point, Inc.
2005 Marietta Rd., NW
Atlanta, GA 30318
(mail order)

Fypon, Inc.
P.O. Box 365
Stewartstown, PA 17363

Klise Mfg. Co.
601 Maryland Ave., NE
Grand Rapids, MI 49505

Old World Mouldings & Finishings
115 Allen Blvd.
Farmingdale, NY 11735

Paints, Stains and Finishes

The companies listed below are manufacturers and generally supply their products only to distributors such as hardware stores, paint and wallpaper stores, and building materials suppliers. If you cannot find the product you want locally, contact the manufacturer for the name of the nearest distributor or order the product from one of the mail-order woodworkers' supply companies listed under "Tools."

Samuel Cabot, Inc.
1 Union St.
Boston, MA 02108

Daly's Woodfinishing Products
1121 N. 36th St.
Seattle, WA 98103

Deft, Inc.
17451 Von Karmen Ave.
Irvine, CA 92714

The Flecto Co., Inc.
Flecto International, Ltd.
P.O. Box 12955
Oakland, CA 94608

Formby's, Inc.
P.O. Box 667
Olive Branch, MS 38654

Glidden Coatings & Resins
Div. of SCM Corp.
900 Union Commerce Bldg.
Cleveland, OH 44115

Hope Co., Inc.
2052 Congressional Dr.
St. Louis, MO 63141

McCloskey Varnish Co.
7600 State Rd.
Philadelphia, PA 19136

Minwax Co., Inc.
102 Chestnut Ridge Plaza
Montvale, NJ 07645

Benjamin Moore & Co.
Chestnut Ridge Rd.
Montvale, NJ 07645

Watco Dennis Corp.
Michigan Ave. at 22nd St.
Santa Monica, CA 90404

Manufactured Door Hardware

The following companies include manufacturers who supply door hardware to large and small hardware stores nationwide, and mail-order retail outlets.

Baldwin Hardware Mfg. Corp.
841 Wyomissing Blvd.
P.O. Box 82
Reading, PA 19603

Ball & Ball
463 W. Lincoln Hwy.
Exton, PA 19341
(mail order)

Kwikset
Div. of Emhart Industries, Inc.
Anaheim, CA 92803

McKinney Forged Iron Hardware
820 Davis St.
Scranton, PA 18505

Omnia Industries, Inc.
P.O. Box 263
49 Park St.
Montclair, NJ 07042

Renovator's Supply Co., Inc.
71 Northfield Rd.
Millers Falls, MA 01349
(mail order)

Ritter & Son Hardware
46901 Fish Rock Rd.
Gualala, CA 95445
(mail order)

Schlage Lock Co.
2401 Bayshore Blvd.
P.O. Box 3324
San Francisco, CA 94119

Handwrought and Handforged Door Hardware

The following list includes both forges staffed by several metalsmiths and by individual craftsmen. Most do some production items as well as custom work. Some, like Newton Millham, specialize in authentic historical reproductions. Others usually make larger-scale items like ornamental fences, gates and window grilles. Ask for a catalog or brochure when you make your inquiry.

Steve Elling
Intarsia
8th Floor
740 Broadway
New York, NY 10003

Dimitri Gerakaris
Upper Gates Rd.
R.F.D. #2
North Canaan, NH 03741

Gregg Leavitt
Upper Bank Forge
Valleybrook Rd.
Wawa, PA 19063

Thomas C. Maiorana
R.D. #4, P.O. Box 60
Montague, NJ 07827

Newton Millham
672 Drift Rd.
Westport, MA 02790

Williamsburg Blacksmiths, Inc.
Buttonshop Rd.
Williamsburg, MA 01096

Worthington Forge
170 S. Main St.
Yardley, PA 19067

Manufactured Window Hardware

Blaine Window Hardware, Inc.
1919 Blaine Dr.
R.D. #4
Hagerstown, MD 21740
(mail order)

Quaker City Manufacturing
701 Chester Pike
Sharon Hill, PA 19079
(mail order)

Selby Furniture Hardware Co., Inc.
17 E. 22nd St.
New York, NY 10010
(mail order)

Glass and Glassworking Materials

The companies listed below sell glass, lead came, copper foil, glass cutters, soldering guns, and other materials and tools for making leaded, beveled, etched, engraved and stained glass. They will ship stock items that are mail ordered.

S. A. Bendheim Co., Inc.
122 Hudson St.
New York, NY 10013

Boulder Art Glass Co., Inc.
1920 Arapahoe Ave.
Boulder, CO 80302

Franklin Art Glass Studios, Inc.
222 E. Sycamore St.
Columbus, OH 43206

Jennifer's Glassworks, Inc.
2410 Piedmont Rd., NE
Atlanta, GA 30324

C. R. Laurence Co., Inc.
P.O. Box 21345
Los Angeles, CA 90021

Occidental Art Glass
410 Occidental St.
Seattle, WA 98104

Sunrise Art Glass Studio
1113 Chicago Ave.
Oak Park, IL 60302

Whittemore-Durgin Glass Co.
P.O. Box 2065
Hanover, MA 02339

Art Glass

Check with your local stained glass shop or hobby shop for art glass before contacting these companies for the name of the distributor nearest you.

Bevels Ltd. (stock bevel pieces)
4676 Admiralty Way, Suite 719
Marina Del Rey, CA 90291

Blenko Glass Co. (antique glass)
P.O. Box 67
Milton, WV 25541

Chicago Art Glass, Inc. (hand-ladled sheet glass and rolled glass)
2382 United Ln.
Elk Grove Village, IL 60007

Spectrum Glass Co., Inc. (rolled glass)
24305 Woodinville-Snohomish Hwy.
Woodinville, WA 98072

Unique Wholesale Stock Bevels (stock bevel pieces)
12422 Memorial Dr.
Houston, TX 77024

Beveling Equipment

See also C. R. Laurence Co., Inc., listed above.

Denver Glass Machinery
1705 S. Pearl
Denver, CO 80210

READING RESOURCES

Books

Architectural Woodwork Quality Standards Guide, Specifications and Quality Certification Program. Arlington, Va.: Architectural Woodwork Institute, 1978.

Ball, John E. *Carpenters and Builders Library, No. 4: Millwork Power Tools, Painting.* Indianapolis: Theodore Audel, 1976.

Bard, Rachel.*Successful Wood Book: How to Choose, Use, and Finish Every Kind of Wood.* Farmington, Mich.: Structures, 1978.

Black & Decker Power Tool Carpentry. New York: Van Nostrand Reinhold, 1978.

Blackburn, Graham J. *Illustrated Interior Carpentry.* Indianapolis: Bobbs-Merrill, 1978.

Boericke, Art and Shapiro, Barry. *Handmade Houses: A Guide to the Woodbutcher's Art.* San Francisco: Scrimshaw Press, 1973.

———. *The Craftsman Builder.* New York: Simon & Schuster, 1977.

Burns, Al. *The Skylight Book.* Philadelphia: Running Press, 1976.

Butler, Robert L. *Wood for Wood-Carvers and Craftsmen.* San Diego: A. S. Barnes, 1975.

Carter, Joe, ed. *Solarizing Your Present Home: Practical Solar Heating Systems You Can Build.* Emmaus, Pa., Rodale Press, 1981.

Castle, Wendell and Edman, David. *Wendell Castle Book of Wood Laminations.* New York: Van Nostrand Reinhold, 1980.

Clery, Val. *Doors.* New York: Viking Press, 1978.

———. *Windows: A Feast for the Eye and the Imagination.* New York: Viking Press, 1978.

Conran, Terence. *The House Book.* New York: Crown, 1974.

Cruz, William. *Will's Guide to Building the $9,000 House.* Santa Cruz, Calif.: Unity Press, 1978.

Cummings, Abbott Lowell. *The Framed Houses of Massachusetts Bay, 1625–1725.* Cambridge, Mass.: Harvard University Press, 1979.

Davidson, Marshall B. *The American Heritage History of Notable American Houses.* New York: American Heritage, 1971.

Dennis, Ben and Case, Betsy. *Houseboat: Reflections of North America's Floating Homes . . . History, Architecture, and Lifestyles.* Seattle: Smuggler's Cove, 1977.

Doors & Windows. Alexandria, Va.: Time-Life, 1978.

The Early American Society Sourcebook. Gettysburg, Pa.: The Early American Society, 1979.

Editors of Consumer Guide. *The Tool Catalog: An Expert Selection of the World's Finest Tools.* New York: Harper & Row, 1978.

———. *Whole House Catalog.* New York: Simon & Schuster, 1976.

Ehrlich, Jeffrey and Mannheimer, Marc. *The Carpenter's Manifesto: A Total Guide that Takes All the Mystery Out of Carpentry for Everybody.* New York: Holt, Rinehart & Winston, 1977.

Feirer, J. L. *Cabinetmaking & Millwork,* rev. ed. New York: Charles Scribner's Sons, 1977.

Fine Woodworking Biennial Design Book. Newtown, Conn.: Taunton Press, 1977.

Fine Woodworking Design Book Two. Newtown, Conn.: Taunton Press, 1979.

Fine Woodworking Techniques, No. 1. Newtown, Conn.: Taunton Press, 1978.

Fine Woodworking Techniques, No. 2. Newtown, Conn.: Taunton Press, 1980.

Forest Products Laboratory, Forest Service, U. S. Department of Agriculture. *Durability of Exterior Natural Wood Finishes in the Pacific Northwest* (Research Paper FPL366). Washington, D.C.: U. S. Government Printing Office, 1980.

Forest Products Laboratory, Forest Service, U.S. Department of Agriculture. "Painting and Finishing." In *The Wood Handbook*, rev. ed. (Agriculture Handbook No. 72, 001-000-03200-3) pp. 16-1-16-13. Washington, D.C.: U. S. Government Printing Office, 1974.

French, Jennie. *Glass-Works: The Copper Foil Technique of Stained Glass.* London: Litton, 1974.

Gault, Lila and Weiss, Jeffrey. *Small Houses.* New York: Warner, 1980.

Gibbia, S. W. *Wood Finishing and Refinishing.* New York: Van Nostrand Reinhold, 1971.

Gick, James. *Creating with Stained Glass.* Laguna Hills, Calif.: Gick, 1976.

Hand, Jackson. *How to Do Your Own Wood Finishing.* New York: Harper & Row, 1976.

Haney, Robert and Ballantine, David. *Wood-stock Handmade Houses.* New York: Random House, 1974.

Hoadley, R. Bruce. *Understanding Wood: A Craftsman's Guide to Wood Technology.* Newtown, Conn.: Taunton Press, 1980.

Hotton, Peter. *So You Want to Fix Up an Old House.* Boston: Little, Brown, 1979.

Isenberg, Anita; Isenberg, Seymour; and Millard, Richard. *Stained Glass Painting.* Radnor, Pa.: Chilton, 1979.

Jackson, Albert and Day, David. *Tools and How to Use Them: An Illustrated Encyclopedia.* New York: Alfred A. Knopf, 1978.

Kern, Ken; Kogon, Ted; and Thallon, Rob. *The Owner-Builder and the Code: Politics of Building Your Home.* Oakhurst, Calif.: Owner-Builder Publications, 1976.

Koch, Robert. *Louis C. Tiffany, Rebel in Glass.* New York: Crown, 1966.

Krenov, James. *A Cabinetmaker's Notebook.* New York: Van Nostrand Reinhold, 1976.

——. *The Fine Art of Cabinetmaking.* New York: Van Nostrand Reinhold, 1977.

——. *The Impractical Cabinetmaker.* New York: Van Nostrand Reinhold, 1979.

Langdon, William K. *Movable Insulation.* Emmaus, Pa.: Rodale Press, 1980.

Lidz, Jane. *Rolling Homes: Handmade Houses on Wheels.* New York: A & W Visual Library, 1979.

Lucie-Smith, Edward. *The Story of Craft: The Craftsman's Role in Society.* Ithaca, N.Y.: Cornell University Press, 1981.

Mariacher, Giovanni. *Glass from Antiquity to the Renaissance.* Translated by Michael Cunningham. London: Hamlyn, 1970.

Mollica, Peter. *Stained Glass Primer, Vol. 1.* Oakland, Calif.: Mollica Stained Glass Press, 1973.

——. *Stained Glass Primer, Vol. 2.* Oakland, Calif.: Mollica Stained Glass Press, 1977.

Mulligan, Charles and Higgs, Jim. *The Wizard's Eye: Visions of American Resource-fulness.* San Francisco, Calif.: Chronicle, 1978.

Needleman, Carla. *The Work of Craft: An Inquiry into the Nature of Crafts and Crafts-manship.* New York: Alfred A. Knopf, 1979.

Norman, Barbara. *Engraving and Decorating Glass.* New York: McGraw-Hill, 1972.

Pain. F. *The Practical Wood Turner.* New York: Sterling, 1979.

Practical Wood Finishing Methods. Pittsburgh, Pa.: Rockwell International, 1978.

Radford, Penny. *Rooms for Living.* London: Design Council, 1976.

Raglan, Lord. *The Temple and the House.* New York: W. W. Norton, 1964.

Reader's Digest Complete Do-It-Yourself Manual. Pleasantville, N. Y.: Reader's Digest, 1973.

Rifkind, Carole. *A Field Guide to American Architecture.* New York: New American Library, 1980.

Rigan, Otto B. *New Glass.* Westminster, Md.: Ballantine, 1977.

Robertson, R. A. *Chats on Old Glass,* rev. ed. Magnolia, Mass.: Peter Smith.

Savage, George. *Glass.* London: Octopus, 1972.

Schmidt, Fred M. *The Window Book.* Coral, Pa.: Fred M. Schmidt, 1976.

Selbo, M. L. *Adhesive Bonding of Wood.* New York: Sterling, 1978.

Spielman, Patrick. *Making Wood Signs.* New York: Sterling, 1981.

Stokes, Gordon. *Modern Wood Turning.* New York: Sterling, 1979.

Talbot, Anthony. *Handbook of Doormaking, Windowmaking and Staircasing.* New York: Sterling, 1980.

Wade, Alex. *A Design and Construction Handbook for Energy-Saving Houses.* Emmaus, Pa.: Rodale Press, 1980.

———. *Thirty Energy-Efficient Houses . . . You Can Build.* Emmaus, Pa.: Rodale Press, 1977.

Wagner, Willis H. *Modern Carpentry: Building Construction Details in Easy-to-Understand Form.* South Holland, Ill.: Goodheart-Willcox, 1976.

Wampler, Jan. *All Their Own: People and the Places They Build.* New York: Oxford University Press, 1978.

Weiss, Jeffrey. *Outdoor Places.* New York: W. W. Norton, 1980.

———. *Pine.* New York: Harper & Row, 1980.

Weiss, Jeffrey and Wise, Herbert H. *Good Lives.* New York: Quick Fox, 1977.

Wise, Herbert H. *Attention to Detail.* New York: Quick Fox, 1979.

———. *Rooms with a View.* New York: Quick Fox, 1978.

Wise, Herbert H. and Weiss, Jeffrey. *Living Places.* New York: Quick Fox, 1976.

———. *Made with Oak.* New York: Quick Fox, 1975.

Magazines

Fine Homebuilding
The Taunton Press, Inc.
52 Church Hill Rd.
P.O. Box 355
Newtown, CT 06470

Fine Woodworking
The Taunton Press, Inc.
52 Church Hill Rd.
P.O. Box 355
Newtown, CT 06470

Glass Studio Magazine
P.O. Box 23383
Portland, OR 97223

New Shelter
Rodale Press, Inc.
33 E. Minor St.
Emmaus, PA 18049

The Old-House Journal
The Old-House Journal Corp.
69A Seventh Ave.
Brooklyn, NY 11217

Stained Glass
The Stained Glass Assoc. of America
1125 Wilmington Ave.
St. Louis, MO 63111

MAKERS OF HANDCRAFTED DOORS AND WINDOWS

Not all of the craftspeople and companies listed here have work appearing in the pages of *Handcrafted Doors and Windows* (see the Photo and Design Credits for names of those who do). However, all are doormakers and windowmakers we would have liked to feature.

Also consider making inquiries of your architect or builder for the names of fine woodworkers and glass artisans in your area who could carry out a commission to build a handcrafted door or window for you.

Makers of Custom Doors and Windows

The craftspeople listed below generally make one-of-a-kind items. Although very often they are experienced in other areas of architectural woodwork like staircasing and cabinetry, they came to our attention for doormaking (d), windowmaking (w), or both (d&w). Almost all the craftspeople work in wood; some work in other media as well (Bruce Fink, for example, works in wood, metal and plastics; Larry Golden works in wood and acrylics). To get more information about the work of individual craftsmen, write a letter of inquiry. Some have brochures that illustrate their work.

Tom Anderson (d)
Creative Openings
P.O. Box 2566
Bellingham, WA 98227

Steve Ball (d)
Rt. #7 MLC
Tallahassee, FL 32308

Tom Bender (w)
Neahkahnie Mountain
38755 Reed Rd.
Nehalem, OR 97131

Lesta Bertoia (d)
R.D. #1
Barto, PA 19504

Michael Bullard (d)
Rt. #7 MLC
Tallahassee, FL 32308

The Cascades Woodwrights (d)
1319½ E. Tennessee St.
Tallahassee, FL 32303

David Chek (d)
1827 Summit Pl. NW #301
Washington, DC 20009

Aj Darby (d)
7900 St. Helena Rd.
Santa Rosa, CA 95404

William Davidson (d&w)
Apache Canyon Woodworks
Rt. #3, Box 94-I
Santa Fe, NM 87501

Volker de la Harpe (d)
P.O. Box 641
Santa Fe, NM 87501

Bruce Fink (d&w)
Rt. #1, Box 278
Woodstock, CT 06281

Al Garvey (d)
281 Scenic Rd.
Fairfax, CA 94930

Larry Golden (d&w)
Rt. #2
Deronda, WI 54008

Peter Good (d)
1966 Tiffin Rd.
Oakland, CA 94602

David Haust (d)
R.D., Box 11A
East Chatham, NY 12060

Adi Hienzsch (d)
Edelweiss Woodcarving Studio
14410 436th SE
North Bend, WA 98045

Victor Hiles (d)
Box 1109
Homer, AK 99603

David L. Hofstad (d)
631 N. Park
Reedsburg, WI 53959

Bob Jepperson (d)
3358 Mosquito Lake Rd.
Deming, WA 98244

Bill Johnson (d)
Denman Island, BC V0R 1T0
Canada

Karen Ann Moulton (d)
Box 51
Marshfield, VT 05658

Saswathan Quinn (d)
1515 Sherwin Ave.
Emeryville, CA 94608

Bob Richardson (d)
Wood Design
P.O. Box 775
Santa Fe, NM 87501

James Schriber (d)
P.O. Box 1145
New Milford, CT 06776

Vic Schulman (d)
Pickles Rd.
Denman Island, BC V0R 1T0
Canada

Israel Serr (d)
Rt. #4, Box 85-D
Santa Fe, NM 87501

Charlie Southard (d)
Box 5212
Santa Fe, NM 87502

Union Woodworks (d)
7 Belknap St.
Northfield, VT 05663

Chuck Weaver (d)
First Impressions
2600 Terry Lake Rd.
Fort Collins, CO 80524

John Zoltai Studio (d)
Box 31
El Rito, NM 87530

Makers of Stock Doors and Windows

Although all of the companies listed below will make custom doors and windows on request, their emphasis is on production items. Donald Pecora's company, Entrances, Inc., specializes in making insulated wooden doors. Other companies, like Maurer & Shepherd Joyners, Inc. and Moser Brothers, make historical reproductions.

Donald Pecora (d)
Entrances, Inc.
Route 123
Alstead, NH 03602

Drums Sash & Door Company
   (d&w)
P.O. Box 207
Drums, PA 18222

Marin Lumber, Inc. (d&w)
15 Jordan St.
San Rafael, CA 94915

Maurer & Shepherd Joyners, Inc.
   (d&w)
122 Naubuc Ave.
Glastonbury, CT 06033

Moser Brothers (d&w)
3rd & Green Sts.
Bridgeport, PA 19405

Renovation Concepts, Inc. (d&w)
P.O. Box 3720
Minneapolis, MN 55403

Custom Builders

The following custom builders make handcrafted doors and windows as part of the houses they build. If you would like to build a house incorporating such architectural details, feel free to contact them. If you only want to commission an individual door or window, refer to the lists of custom and stock doormakers and windowmakers above.

Stan Griskivich
Energy Efficient Hand Made Houses
153B Cousins Island, R.R. #1
Yarmouth, ME 04096

Hank Huber
R.F.D. #2, Box 371
Peterborough, NH 03458

Hagen & Sons Construction
95 Brodin Ln.
Watsonville, CA 95076

T. Larsen
Denman Island, BC V0R 1T0
Canada

James Parker
Cerro Construction
1880 Forest Cir.
Santa Fe, NM 87501

Makers of Art Glass for Doors and Windows

Stained, beveled and etched glass makers are included in the list below.

Brusey Stained Glass Studio
2510 S. Main
Soquel, CA 95073

Joann Chamberlain
Rt. #7 MLC
Tallahassee, FL 32308

Rhonda L. Dixon
General Delivery
La Rue, AR 72743

Jim & Deborah Hannen
Cannon Beach Stained Glass
987 S. Hemlock
Cannon Beach, OR 97110

Mark Hattman
535 Fairchild Ave.
Kent, OH 44240

Adele Hiles
Box 1109
Homer, AK 99603

Rachel Josepher-Gaspers
P.O. Box 633
Bolinas, CA 94924

Betty J. Kilpatrick
247 Rim Rock Dr.
Durango, CO 81301

Lynn C. Kraft
2083 Main St.
Box 73
Center Valley, PA 18034

Cheer Owens
P.O. Box 1088
St. George, UT 84770

Beverly Rieser
6979 Exeter Dr.
Oakland, CA 94611

Ted Ricchiuti
T.D.R. Studios
P.O. Box 296
Seaside Park, NJ 08752

Bruce Sherman
4071 23rd St.
San Francisco, CA 94114

Standing Rock Designery Stained
   Glass Studio
715 North Mantua St.
Kent, OH 44240

Megan Timothy
Designer Glassworks
11159 La Maida St.
North Hollywood, CA 91601

Mark Stine
Transparent Dreams
1350 Cook St.
Denver, CO 80206

Lonnie Willis
Glass Studio
Down Industrial Park
Railroad Ave.
Tallahassee, FL 32301

Carla Wingate
P.O. Box 883
Stinson Beach, CA 94970

PHOTO AND DESIGN CREDITS

To determine the credits for a particular photograph, locate the page and position in which it appears in the left-hand column. The name of the photographer, and then the craftsman or supplier, appears opposite the page number in the right-hand column.

ii, opposite title	W. R. Eckerman; Betty Kilpatrick
vi	Don Arns; Per Hojestad
vii, left	B. H. Zoss; craftsman unknown
vii, right	Tom Bender; craftsman unknown
viii, left	Margaret Smyser; craftsman unknown
viii, right	Timothy J. Krohn; David L. Hofstad
ix	Mitch Mandel; Carla Wingate (window); Didrik Pederson (door)
xii	Mory Shand; Val Heckrodt
xiii, top and bottom	B. H. Zoss; craftsman unknown
xiv	Tom Bender; craftsman unknown
xv, top left	Tom Bender; craftsman unknown
xv, center left	B. H. Zoss; craftsman unknown
xv, bottom left	Tom Bender; craftsman unknown
xv, top right	B. H. Zoss; craftsman unknown
xvi	Tom Bender; J. B. Blonk
xvii, left	Paul Bailey; Tom Larsen
xvii, right	Paul Bailey; Bill Johnson
xviii	Mitch Mandel; craftsman unknown
xix, top left	James Warfield; craftsman unknown
xix, top right	Tom Bender; craftsman unknown
xix, bottom	Tom Bender; craftsman unknown
xx, top, center and bottom	Tom Bender; craftsman unknown
xxi	Tom Bender; craftsman unknown
xxii, top	Tom Bender; craftsman unknown
xxii, bottom	B. H. Zoss; craftsman unknown
xxiii	Bill French; Cecil and Dorothy Brusey
xxiv and 1	Bud Lewis; Deck Dargan
2	Mitch Mandel; John Peck
4, top	Mory Shand; Val Heckrodt
4, bottom	Peter Leach; Larry Golden

6	Fritz Rothermel; craftsmen unknown; photos courtesy of the Planning Commission of the City of Reading, Pennsylvania
9	Carl Doney; Larry Golden
10	Carl Doney; reconstituted veneer samples courtesy of the Dean Company; plain veneers from Albert Constantine & Son, Inc.
12	Laura Chek; David Chek
13, top	Bruce Fink; Bruce Fink
13, bottom	Roy Mullin; Victor Hiles (door); Adele Hiles (window)
15	Chris Barone; stock glues courtesy of Wentz Hardware
16, left	Carl Doney; Bob Jepperson
16, center	Roy Mullin; Fritz Grant
16, right	Margaret Smyser; Bob LeMay
19, top	Sally Ann Shenk; Charlie Southard
19, center	Bill French; Robert Sommerville
19, bottom	Sally Ann Shenk; William Davidson
20, left	Michael Kanouff; Peter Good
20, center	Carl Doney; Bob Jepperson
20, right	Sally Ann Shenk; John Zoltai
24, left	Clint Crawley; Beehive Glass
24, center	Michael Kanouff; Jim Vandegrift
24, right	Stan Griskivich; Stan Griskivich
26, left	Lois Moulton; Karen Ann Moulton
26, center	Mitch Mandel; Donald N. Solow, Arch.
26, right	Mitch Mandel; Tony Barbera
27	T. L. Gettings; Larry Golden
31, left	Roy Mullin; Victor Hiles
31, right	Mitch Mandel; Gary Geyer
32, left	Mitch Mandel; Gary Geyer
32, center	Roy Mullin; Jim Shoppert
32, right	Mitch Mandel; Steve Meachum
33	T. L. Gettings; Larry Golden
34	Mitch Mandel; Harold Horowitz
36	Carl Doney; molding samples courtesy of Driwood Mouldings; handmade moldings by Tom Walz of Rodale Press
38	Chris Barone; door by Phil Gehret of Rodale Press; inlay from The Woodworkers' Store
39	Carl Doney; inlay from The Woodworkers' Store
41, top	Sally Ann Shenk; Israel Serr
41, bottom	Carl Doney; Glid-Tone stains courtesy of Glidden Coatings & Resins, Division of SCM Corporation; Minwax stains courtesy of Minwax
42	Sally Ann Shenk; Charlie Southard
43	Otto Rigan; James Hubbell
44	Carl Doney; forged iron hinges courtesy of Newton Millham; brass hinges from Omnia Industries
45	Paul Bailey; Tom Larsen (door); Ray Lipovski (hinges)
47, left	James Schriber; James Schriber
47, center	Earl A. Jett/Jettcraft; craftsman unknown
47, right	John Elliott; David Haust (door); Thomas C. Maiorana (latch)

48, top	Paul Bailey; Tom Larsen
48, bottom left	Michael Kanouff; Bruce Reeves
48, bottom right	T. L. Gettings; Alexander Weygers
50 and 51	Carl Doney; all Baldwin and Schlage hardware courtesy of Allen Hardware; remaining items courtesy of Renovator's Supply, Inc.
53	Carl Doney; lion-head door knocker courtesy of Ball & Ball Hardware; remaining items courtesy of Allen Hardware
54 and 55	Robin Rothstein; Tony Barbera
56 and 58	Glenn Sharon; Tom Barr
60, 61, 63, 64 and 65	Carl Doney; Bob Jepperson
66 and 69	Mitch Mandel; Aj Darby
70, 72 and 73	Ira Gavrin; Donald Pecora
74, 76 and 79	B. Edwin Riechert; Adi Hienzsch
80	Sally Ann Shenk; James Parker
81	Sally Ann Shenk; Bob Richardson
82	Sally Ann Shenk; James Parker
85	Sally Ann Shenk; Bob Richardson
86, 88 and 89	Carl Doney; Lesta Bertoia
90	John Hamel; Tom Anderson
93, top left and center	Carl Doney; Tom Anderson
93, top right and bottom	John Hamel; Tom Anderson
94, 95 and 96	T. L. Gettings; Alexander Weygers
97	Michael Kanouff; Alexander Weygers
98, 100, 101 and 103	Michael Kanouff; Al Garvey
104 and 105	Mitch Mandel; Moser Brothers, Inc. (Craftsman shown in inset is John Derfler of Moser Brothers, Inc.)
106	Tom Bender; Goddard College students
107, top	B. H. Zoss; craftsman unknown
107, bottom	B. H. Zoss; Lothrop Merry
109, top left	Hank Huber; Hank Huber
109, top right and bottom	Mitch Mandel; Stan Saran
111	Stan Griskivich; Stan Griskivich
113, top left	Rob Super; Beverly Reiser
113, top right	Joseph A. Gaspers; Rachel Josepher-Gaspers
113, bottom	Robin Rothstein; Tom Watson
116	Bill French; Hagen & Sons
117, top left	Mitch Mandel; Blackmun Builders
117, top right	Tom Bender; craftsman unknown
117, bottom	Michael Kanouff; craftsman unknown
120	Helen Schwartz; craftsman unknown
126	B. H. Zoss; Lothrop Merry
128, top	Mitch Mandel; craftsman unknown
128, center and bottom	Christie Tito; craftsman unknown
129	Mitch Mandel; Rich Kline
131	Michael Kanouff; craftsman unknown
132	Hal Denison; James Hannen of Cannon Beach Stained Glass
134, top	Luke Golobitsch; craftsman unknown
134, bottom	Mitch Mandel; Jerry Cebe

135, top left	Jim MacInnes; Beverly Brecht
135, top right	Clint Crawley; Cheer Owens
135, center	Mitch Mandel; Carla Wingate
135, bottom	Mitch Mandel; Jerry Cebe
136, top	Shirley Sanford; Shirley Sanford
136, bottom	Mitch Mandel; Bill and Sandia Bruton
137, top	Carita Parker; John Herring, designer; Kay Herring and Earl McFarland, craftspeople, all of Standing Rock Designery Stained Glass Studio
137, center	Mitch Mandel; Leona McNeil
137, bottom	Rob Super; Beverly Reiser
138	B. H. Zoss; Steve Gennadli
140	Rhonda Dixon; Rhonda Dixon
141, left	Bill French; Lenore Glaum
141, right	Mitch Mandel; Lenore Glaum
142 and 143	Craig Du Monte; Rachel Josepher-Gaspers
144, 146, 147 and 148	Carl Doney; Larry Golden
150, 154 and 155	Mitch Mandel; Bruce Fink
156, 158, 159, 160, 161 and 163	John Hamel; Megan Timothy
164	Bruce Sherman; Bruce Sherman
165, 166 and 169	Michael Kanouff; Bruce Sherman
170 and 172	John Hamel; Lynn Kraft
173	Lynn Kraft; Lynn Kraft
175	John Hamel; Lynn Kraft
176 and 178	Carl Doney; Joe Devlin
179, left	Mitch Mandel; Gregg Leavitt
179, right	John Hamel; Jack Boyd
180, 182 and 183	Carl Doney; Tom Bender
184	Robin Rothstein; David Adams

INDEX

T

Thompkins Watersealer, 59
Threshold, of doorway, 29
Thumblatch, for doors, 47
Tools
    for panel doors, 20
    safe use of, 14-15
    for windowmaking, 118
Trim. *See* Moldings
Tung oil, used as finish, 41

V

Varnish, for door finish, 14
Veneers
    types of, 8-10
    use of, 7

W

Walnut
    used for doors, 14
    used for veneer, 9
Warm Door, 70-73
Weldwood plastic resin glue, 57

Window grates, for security, 176-79
Windows. *See also* Glasswork
    accessories for, 114
    building codes and, 108
    casement, 123-27
    channels for, 128
    construction of, 116-31
    construction standards for, 112
    double-glazed, 109
    double-hung, 128-30
    energy efficiency of, 109-10
    glazing materials for, 110
    lumber for, 117-18
    preassembled, 112-14
    skylights, 130-31
    structure of, 106-15
    styles of, 107-8, 116-17
    tools for, 108
    weather stripping for, 127
Window wall, roll-away, 180-83
Wood filler, use of, 39-40
Woods
    for doors, 14
    for handles, 65
    for veneer, 9
    for windows, 117-18